Firefighter Fatalities in the United States in 1998

Prepared for

United States Fire Administration
Federal Emergency Management Agency
Contract No. EME-1998-CO-0202-T0004

Prepared by

IOCAD Emergency Services Group

August 1999

Table of Contents

ACKNOWLEDGMENTS

This study of firefighter fatalities would not have been possible without the cooperation and assistance of many members of the fire service across the United States. Members of individual fire departments, chief fire officers, the National Interagency Fire Center, US Forest Service personnel, the US military, the Department of Justice, the National Fire Protection Association, and many others contributed important information for this report.

IOCAD Emergency Services Group of Emmitsburg, Maryland (a division of IOCAD Engineering Services, Inc.) conducted this analysis for the United States Fire Administration under contract EME-1998-CO-0202-T0004.

The ultimate objective of this effort is to reduce the number of firefighter deaths through an increased awareness and understanding of their causes and how they can be prevented. Fire fighting, rescue, and other types of emergency operations are essential activities in an inherently dangerous profession, and unfortunate tragedies occur. This is the risk all firefighters accept every time they respond to an emergency incident. However, the risk can be greatly reduced through efforts to increase firefighter health and safety.

The United States Fire Administration would like to extend its thanks to the following individuals for providing photographs for the cover of this report:

> Master Firefighter Martin Grube with the Virginia Beach Fire Department.
> Bill Tompkins with the New Jersey Metro Fire Photographers Association.
> Orlando Dominguez with the Brevard County Fire/Rescue.
> G. Emas with the Edmonton Emergency Response Department.

This report is dedicated to the families of those firefighters who made the ultimate sacrifice in 1998. May the lessons learned from their passing not go unheeded.

BACKGROUND

For over 20 years, the United States Fire Administration (USFA) has tracked the number of firefighter fatalities and conducted an annual analysis. Through the collection of information on the causes of firefighter deaths, the USFA is able to focus on specific problems and direct efforts towards finding solutions to reduce the number of firefighter fatalities in the future. This information is also used to measure the effectiveness of current programs directed toward firefighter health and safety.

In addition to the analysis, the USFA maintains a list of firefighter fatalities for the Fallen Firefighters Memorial Service. The fallen firefighter's next of kin, as well as members of the individual fire department, are invited to the annual Fallen Firefighters Memorial Service, which is held at the National Emergency Training Center in Emmitsburg, Maryland every fall. Additional information regarding the memorial service can be found on the internet at **http://www.usfa.fema.gov/ffmem/service.htm** or by calling the National Fallen Firefighters Foundation at (301)-447-1365. An updated list of firefighter fatalities from 1981, through the present, including a searchable database, can be found at **http://www.usfa.fema.gov/ffmem/ffmem_search.cfm**

INTRODUCTION

This report continues a series of annual studies by the United States Fire Administration (USFA) of on-duty firefighter fatalities in the United States.

The specific objective of this study was to identify all of the on-duty firefighter fatalities that occurred in the United States in 1998, and to analyze the circumstances surrounding each occurrence. The study is intended to help identify approaches that could reduce the number of firefighter deaths in future years.

In addition to the 1998 overall findings, this study includes special analyses on basic safety concepts that have the potential of saving firefighter's lives and a discussion of firefighter health.

Who Is a Firefighter?

For the purpose of this study, the term firefighter covers all members of organized fire departments, including career and volunteer firefighters; full-time public safety officers acting as firefighters; state and federal government fire service personnel; including wildland firefighters; and privately employed firefighters, including employees of contract fire departments and trained members of industrial fire brigades, whether full or part-time. It also includes contract personnel working as firefighters or assigned to work in direct support of fire service organizations.

Under this definition, the study includes not only local and municipal firefighters, but also seasonal and full-time employees of the United States Forest Service, the Bureau of Land Management, the Bureau of Indian Affairs, the Bureau of Fish and Wildlife, the National Park Service, and state wildland agencies. The definition also includes prison inmates serving on fire fighting crews; firefighters employed by other governmental agencies such as the United States Department of Energy; military personnel performing assigned fire suppression activities; and civilian firefighters working at military installations.

What Constitutes an On-Duty Fatality?

On-duty fatalities include any injury or illness sustained while on-duty that proves fatal. The term on-duty refers to being involved in operations at the scene of an emergency, whether it is a fire or non-fire incident; being en route to or returning from an incident; performing other officially assigned duties such as training, maintenance, public education, inspection, investigations, court testimony and fund-raising; and being on-call, under orders, or on stand-by duty, except at the individual's home or place of business.

These fatalities may occur on the fireground, in training, while responding to or returning from alarms, or while performing other duties that support fire service operations.

A fatality may be caused directly by accident or injury, or it may be attributed to an occupational-related fatal illness. A common example of a fatal illness incurred on duty is a heart attack. Fatalities attributed to occupational illnesses would also include a communicable disease contracted while on duty that proved fatal, where the disease could be attributed to a documented occupational exposure.

Accidents that claim the lives of on-duty firefighters are also included in the analysis, whether or not they are directly related to emergency incidents.

Injuries and illnesses are included where death is considerably delayed after the original incident. When the incident and the death occur in different years, the analysis counts the fatality as having occurred in the year that the incident occurred. One firefighter died in 1998 as the result of an injury suffered in 1989. The firefighter died of medical complications that resulted from a fall from a pumper that was responding to an emergency in 1989. Because this death was the result of an incident that occurred prior to 1998, it is counted as 1989 fatality for statistical purposes. This death is not included in the 91 fatalities for 1998 that are analyzed in this report. Since this death occurred in 1998, this firefighter will be included in the 1998 annual Fallen Firefighters Memorial Service at the National Emergency Training Center, and his name will be included on the list of firefighters who died in 1998.

Two other firefighter deaths that occurred in 1998 are not included in this report. Two firefighters died in separate incidents in Puerto Rico during 1998. The circumstances surrounding the deaths of these two firefighters are included, along with all other firefighters, in the appendix of this report. Their deaths, however, are not included in this analysis. The scope of this analysis includes firefighter fatalities where the fire department is based within the fifty states and the District of Columbia.

There is no established mechanism for identifying fatalities that result from illnesses that develop over long periods of time, such as cancer, which may be related to occupational exposure to hazardous materials or products of combustion. It has proven to be very difficult over several years to provide a full evaluation of an occupational illness as a causal factor in firefighter deaths, because of the limitations in the ability to track the exposure of firefighters to toxic hazards, the often delayed long-term effects of such exposures, and the exposures firefighters may receive while off-duty.

Sources of Initial Notification

As an integral part of its ongoing program to collect and analyze fire data, USFA solicits information on firefighter fatalities directly from the fire service and from a wide range of other sources. These sources include the Public Safety Officer's Benefit Program (PSOB) administered by the Department of Justice, the National Institute for Occupational Safety and Health (NIOSH), the Occupational Safety and Health Administration (OSHA), the US military, the National Interagency Fire Center, and other federal agencies.

The USFA receives notification of some deaths directly from fire departments, as well as from fire service organizations such as the International Association of Fire Chiefs (IAFC), the International Association of Fire Fighters (IAFF), the National Fire Protection Association (NFPA), the National Volunteer Fire Council (NVFC), state fire marshals, state training organizations, other state and local organizations, fire service internet sites, and fire service publications. The USFA also keeps track of fatal fire incidents as part of its Major Fire Investigations Project and maintains an ongoing analysis of data from the National Fire Incident Reporting System (NFIRS) for the production of the report "Fire in the United States".

Procedure for Including a Fatality in the Study

In most cases, after notification of a fatal incident, initial telephone contact is made with local authorities by the USFA's contractor to verify the incident, its location and jurisdiction, and the fire department or agency involved. Further information about the deceased firefighter and the incident may be obtained from the Chief of the fire department or his or her designee over the phone or by other data collection forms.

Information that is routinely requested includes NFIRS-1 (incident) and NFIRS-3 (fire service casualty) reports, the fire department's own incident reports and internal investigation reports, copies of death certificates or autopsy results, special investigative reports such as those produced by the USFA or NFPA, police reports, photographs and diagrams, and newspaper or media accounts of the incident.

After obtaining this information, a determination is made as to whether the death qualifies as an on-duty firefighter fatality according to the previously described criteria. The same criteria were used for this study as in previous annual studies. Additional information may be requested, either by follow-up with the fire department directly, from state vital records offices, or other agencies. The determination as to whether a fatality qualifies as an on-duty death for inclusion in this statistical analysis is made by the USFA. The final determination as to whether a fatality qualifies as a line of duty death for inclusion in the National Fallen Firefighters Memorial Service is made by the National Fallen Firefighters Foundation.

1998 FINDINGS

Ninety-one (91) firefighters died while on duty in 1998.[1] This represents a drop of three deaths from 1997, and continues the downward trend from the level of 104 firefighter fatalities in 1994. The total of 91 fatalities is the third lowest number recorded in the 21 years that this data has been collected, and is only the sixth time that the total has been less than 100 fatalities. The lowest years were 1992, with 75 fatalities, and 1993, with 77 fatalities.

This year's total continues the long-term downward trend of reduced fatalities that began in 1979, after a peak of 171 in 1978. The overall trend in firefighter fatalities is down twenty-three percent over the last ten years. However, the rate of reduction in the last five years has slowed to five percent, partly due to the uncharacteristically low number of deaths that occurred in 1992 and 1993 (Figure 1).

Figure 1 - On-Duty Firefighter Fatalities (1977-1998)

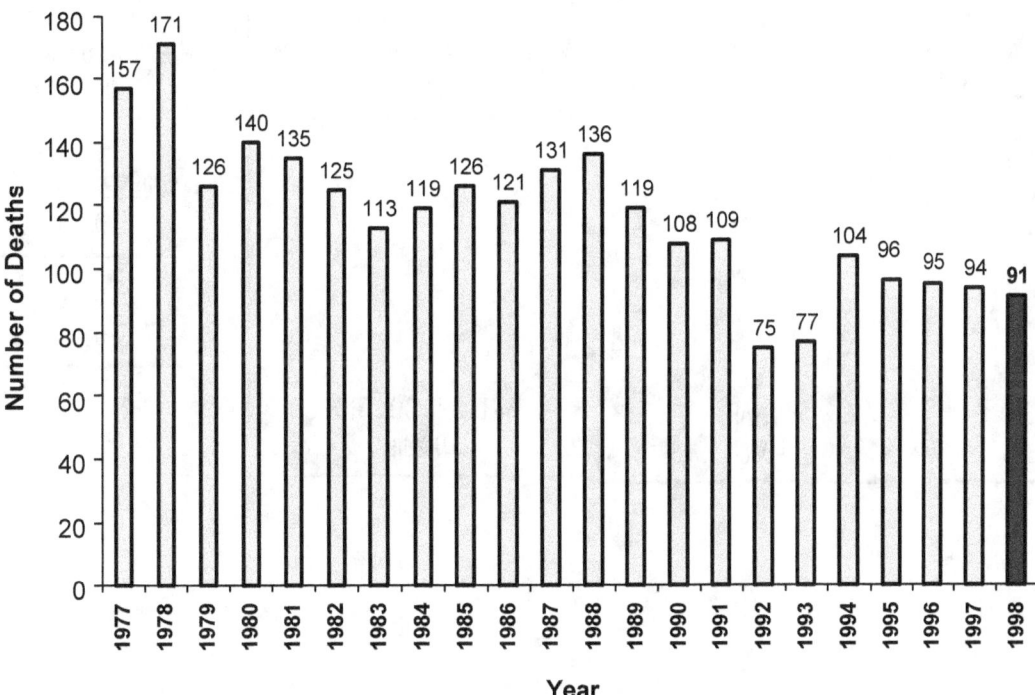

---

[1] As mentioned earlier, the 91 on-duty fatalities in 1998 do not include a firefighter who died in 1998 as the result of injuries he received in 1989 or two Puerto Rican firefighters who died in 1998.

The 1998 firefighter fatalities included 54 volunteer firefighters and 37 career firefighters (Figure 2). Among the volunteer firefighter fatalities, 49 were from local or municipal volunteer fire departments and five were seasonal or contract members of wildland fire agencies. Of the career firefighters who died, 33 were members of local or municipal fire departments, three were wildland firefighters, and one was a member of the Air Force. All ninety-one of the fatalities were men.

Figure 2 – Career vs. Volunteer Deaths

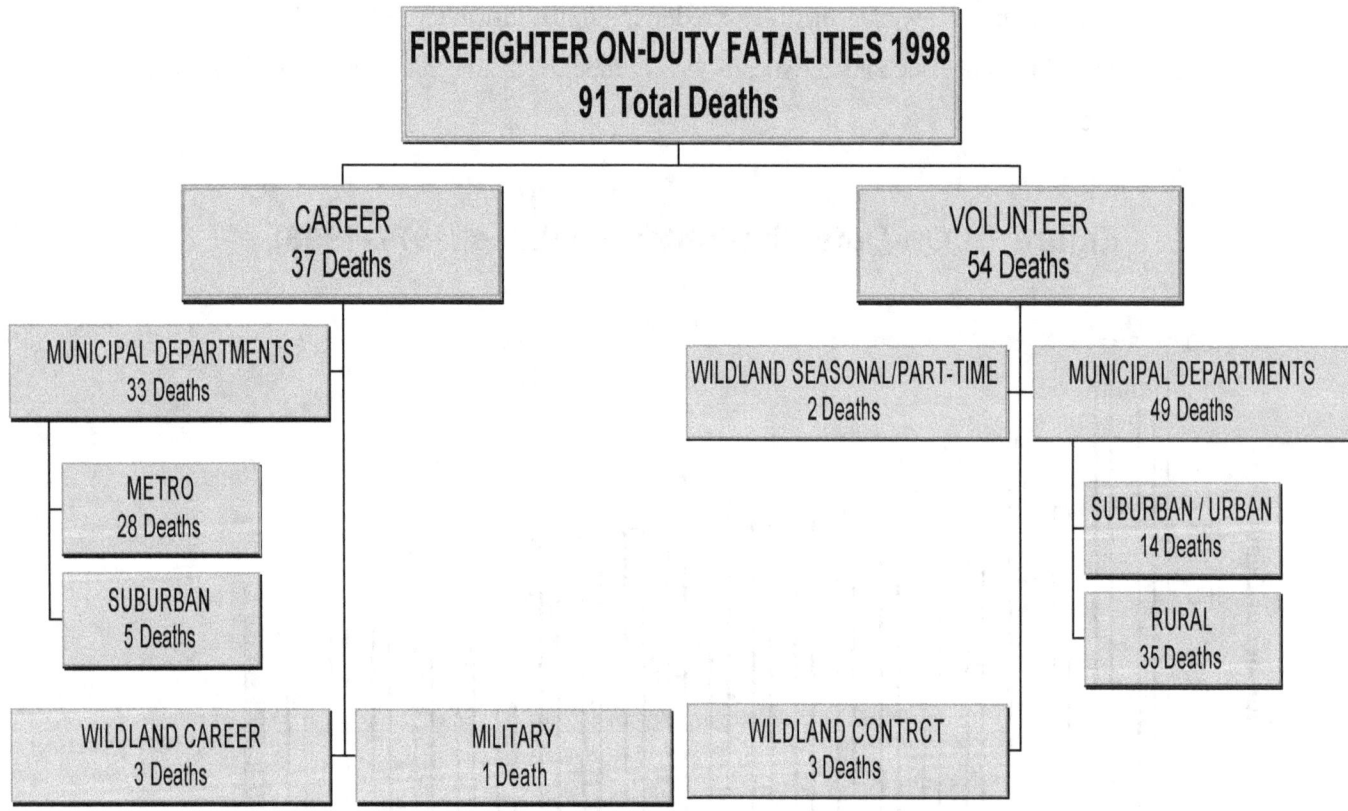

The 91 deaths resulted from 79 incidents. Ten multi-fatality incidents resulted in 22 firefighter deaths. Three California firefighters were killed when their fire department helicopter crashed during a medical transport and three New York firefighters died when they were trapped by rapid fire progress while fighting a high rise apartment building fire. Two Ohio firefighters died when they were trapped by rapid fire progress in a residential basement fire, two Illinois firefighters died when they became disoriented as the result of a backdraft in a commercial building, two Iowa firefighters were killed when they were struck by pieces of a large propane tank that experienced a BLEVE, two New York firefighters were killed as the result of a floor collapse in a commercial/residential occupancy, two firefighters were killed in the crash of an airtanker in New Mexico, two Mississippi firefighters were killed as the result of a roof collapse during a commercial building fire, two firefighters were killed in North Carolina fighting a fire in an auto wrecking yard storage building, and two Kansas firefighters were killed when the ladder that they were repositioning contacted an electric wire and electrocuted both firefighters.

The number of deaths associated with brush, grass or wildland fire fighting rose to thirteen, from the nine deaths experienced in 1997, and the five deaths experienced in 1996. These three years reflect a significant drop from the 18 firefighters that died in wildland activities in 1995. Three firefighters were killed in two wildland fire fighting aircraft crashes in California and New Mexico; two died in separate apparatus collisions in California, one firefighter died while filling tankers (tenders) at a wildfire, one firefighter died when he was involved in a vehicle collision while driving his personal vehicle to a reported wildland fire, one firefighter was killed when the tanker he was driving rolled over while responding to a wildfire, one firefighter was killed while riding the back step of a pumper when he was crushed between the pumper and a utility pole, one firefighter suffered a heart attack after fighting a wildfire for four hours, one firefighter died of burns he received when his bulldozer was overcome by fire progress, one firefighter died of a heart attack that he suffered while protecting exposures at an wildland interface fire, and one firefighter was killed in an unwitnessed rollover while using a bulldozer to maintain a fire road in California.

In 1998, six of the thirteen wildland-related firefighter deaths involved vehicle collisions. One involved a firefighter who was riding the back step of a pumper as it was relocated during a wildfire, one firefighter was killed in a collision involving his personal vehicle while responding to a wildland fire, and a third involved the rollover of a home-built tanker (tender) in Texas. None of these firefighters were wearing a seat belt. One seasonal firefighter and one inmate wildland firefighter were killed in separate collisions involving their wildland fire fighting vehicles. The status of their seat belts were unknown. The seat belt use of the California firefighter killed while maintaining a fire road was also unknown.

Type of Duty

In 1998, 70 firefighter on-duty deaths were associated with emergency incidents, accounting for 77 percent of the 91 fatalities (Figure 3). This includes all firefighters who died while responding to an emergency, while at the emergency scene, or after the emergency incident. Non-emergency activities accounted for 21 fatalities (23 percent). Non-emergency duties include training, administrative activities, or performing other functions that are not related to an emergency incident.

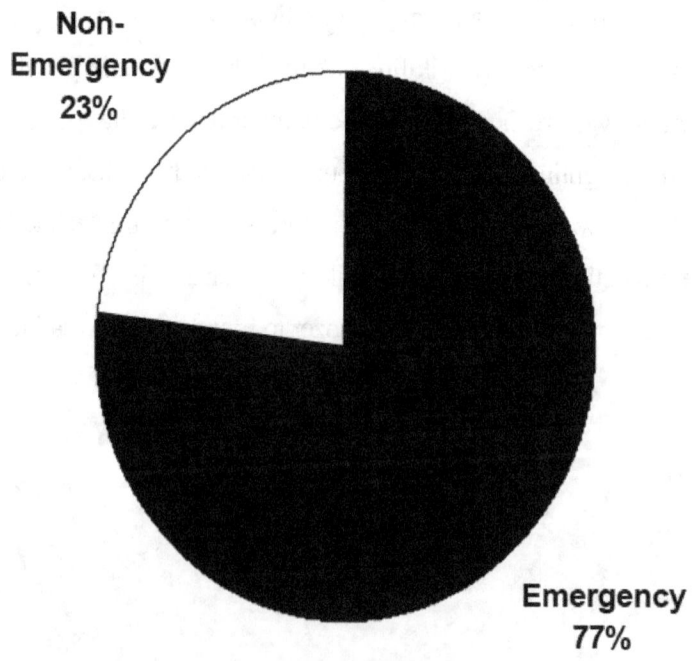

Figure 3 - Firefighter Deaths While Performing Emergency Duty (1998)

Non-Emergency 23%

Emergency 77%

The number of deaths by type of duty being performed is shown in Table 1 and presented graphically in Figure 4. As in previous years, the largest number of deaths occurred during fireground operations. There were 42 fireground deaths, which accounted for 46 percent of the fatalities, up slightly from 44 percent in 1997. Of the 42 fireground deaths, two in five (17) resulted from heart attacks that occurred on the fire scene. Other fireground deaths included 14 from asphyxiation, seven from internal trauma, and two from burns. One firefighter died when he lost too much blood due to a partial leg amputation and one firefighter died as the result of a stroke that occurred after he complained of feeling ill at the scene of a structure fire.

Table 1. Type of Duty – 1998	Number	Percent*
Fireground Operations	42	46%
Responding / Returning from Alarm	14	15%
Non-Fire Emergencies	14	15%
Training	12	13%
Other On-Duty	9	10%
TOTAL	91	99%

Figure 4 - Fatalities by Type of Duty

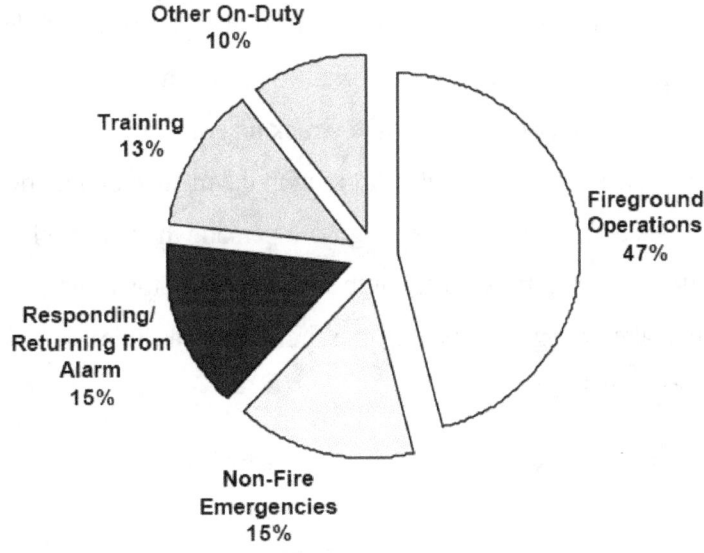

* - **Does not sum to 100% due to rounding.**

Fourteen firefighters died while responding to or returning from emergencies and fourteen died in relation to training activities.

Responding to or returning from emergency incidents has been the second leading type of duty in which firefighter deaths have occurred each year since 1993. There is reason for hope – the fourteen firefighters killed during this activity in 1998 is less than half of the number of deaths that occurred in 1995 when twenty-nine firefighters died. In 1998, ten of these deaths involved volunteer firefighters. Six were killed in motor vehicle collisions – five were killed enroute to the emergency and one was killed while returning to the station. A Fire Police officer died of a heart attack as he arrived at the scene of a motor vehicle collision, one firefighter died while preparing to search for a lost child, and two firefighters suffered heart attacks upon returning to the station. Two career firefighters died while returning from emergencies, both of heart attacks. Two wildland firefighters, one seasonal and one inmate, died in separate vehicle collisions.

Fourteen deaths were related to activities at the scene of non-fire emergency incidents. Five of the deaths involved fire police officers. Three died of heart attacks while directing traffic and two were killed when they were struck by vehicles as they conducted their duties. Three Los Angeles City firefighters were killed when their helicopter crashed during a medical transport, the patient was also killed and two other firefighters were injured. Two firefighters were killed when they were struck by other vehicles while working on the scene of earlier motor vehicle collisions, one firefighter died when a tractor trailer lost control and struck an ambulance and one died when a vehicle lost control on ice and struck a firefighter working in the median of a highway. Two firefighters died of heart attacks, one while searching for a person who was going to commit suicide and one died after providing patient care at the scene of a motor vehicle collision. One firefighter drowned while attempting a rescue and one firefighter was killed as he attempted to remove a fallen tree from a roadway.

Twelve firefighters died during training exercises. This number is significantly higher than has been experienced in the last several years. In 1997, five firefighters died while training, six died in 1996, and the total for 1995, was three. The firefighter deaths in 1998, involved the following circumstances:

- Heart attack after completing annual physical agility test
- Heart attack after completing 30 minutes on a treadmill
- Heart attack during a drill at a home for the elderly
- Heart attack after setting up for live fire training drill
- Heart attack at home after complaining of feeling ill at a drill
- Heart attack while preparing for a CPR class in station
- Heart attack while participating in airport fire fighting training
- Heart attack after completing an SCBA drill
- Fall from a pickup truck while setting up competition training
- Struck by a non-fire department vehicle at the scene of a drill
- Collision on the way to a paramedic training class
- Stroke/Seizure from head injury suffered in training

Nine of the firefighters who died in relation to training exercises were volunteers. Two career firefighters died of heart attacks, one after SCBA training and one during a simulated victim rescue from an aircraft. One career firefighter died of a stroke/seizure some time after bumping his head while completing an SCBA training maze.

There were nine deaths that occurred during non-emergency duty activities. These deaths include three firefighters who died from heart attacks while on duty – one while conducting an inspection, one while in his sleep, and one while the firefighter was working in a light duty assignment in the department's SCBA maintenance shop. Two firefighters were electrocuted as they repositioned a metal ladder for a painting contractor, a firefighter was killed in a vehicle collision as he drove a fire truck to a remote airfield to standby for touch and-go-landings, one firefighter was killed in an unwitnessed tractor rollover as he performed maintenance on a fire road, one firefighter died of a CVA while on duty, and one firefighter died when a blood clot lodged in his lung.

Career, Volunteer, and Wildland Deaths by Type of Duty

Figure 4a depicts career, volunteer, and wildland firefighter deaths by type of duty. Wildland career, wildland seasonal, and wildland contractor deaths were grouped together. This chart demonstrates the disproportionate number of fatalities experienced by volunteer firefighters responding to and returning from alarms when compared with career and wildland firefighters. The large number of career firefighter deaths while on-duty but not involved in incident or training activity may be attributed to the fact that career firefighters are on-duty for longer periods of time than volunteer firefighters. The on-duty periods for volunteer firefighters are generally related to an emergency incident or other official functions such as training. Some volunteer fire departments staff stations overnight similar to a career department but their number is small when compared to the total number of volunteer fire departments.

Figure 4a - Career, Volunteer, and Wildland Deaths by Type of Duty (1998)

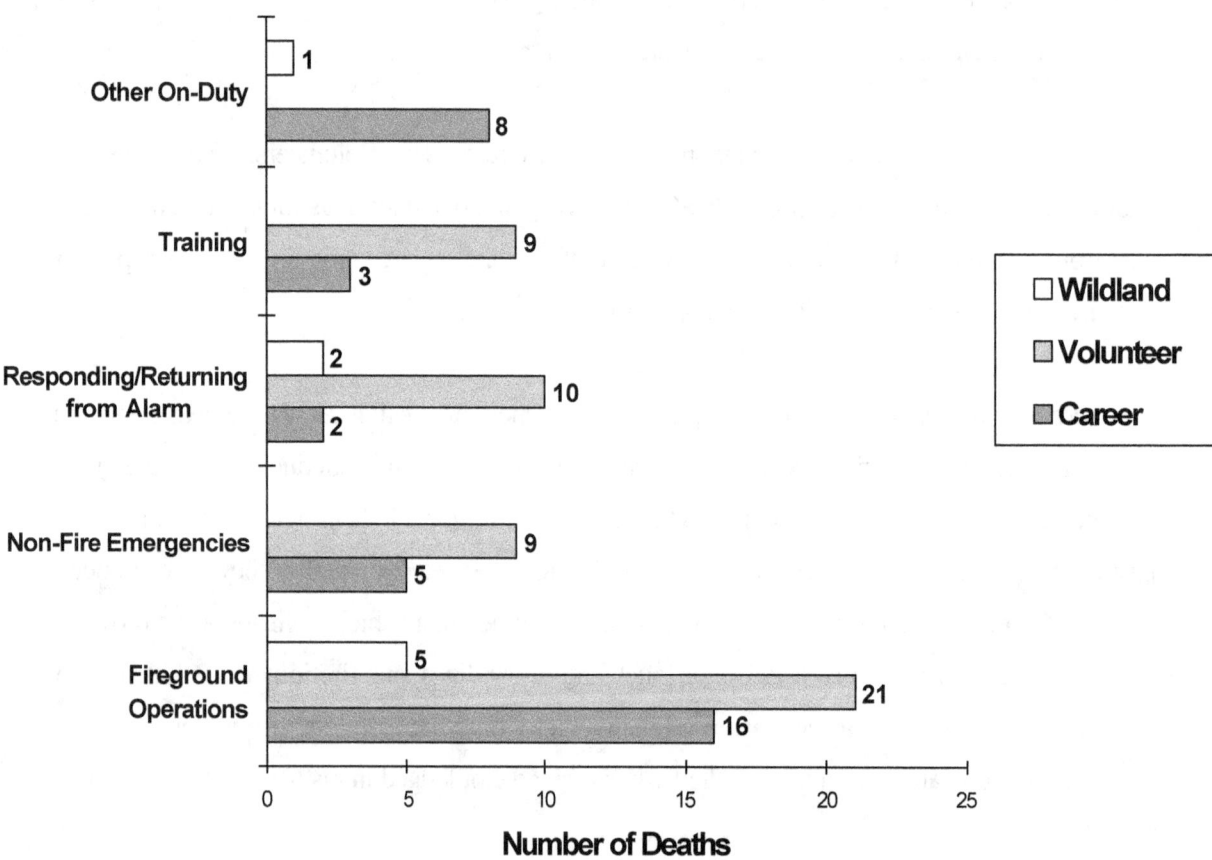

Type of Emergency Duty

65 firefighters died while directly engaged in emergency service delivery, including deaths that were the result of injuries sustained on the incident scene or enroute to the incident scene. Figure 4b shows the percentage of firefighters killed in fire fighting, emergency medical services, hazardous materials, and technical rescue related incidents. 46 firefighters were killed in relation to fires, 14 were killed in relation to EMS calls, two were killed in association with hazardous materials emergencies, one was killed while engaged in a technical rescue, and two firefighters died while preparing for or participating in searches for lost persons.

Figure 4b – Type of Emergency Duty

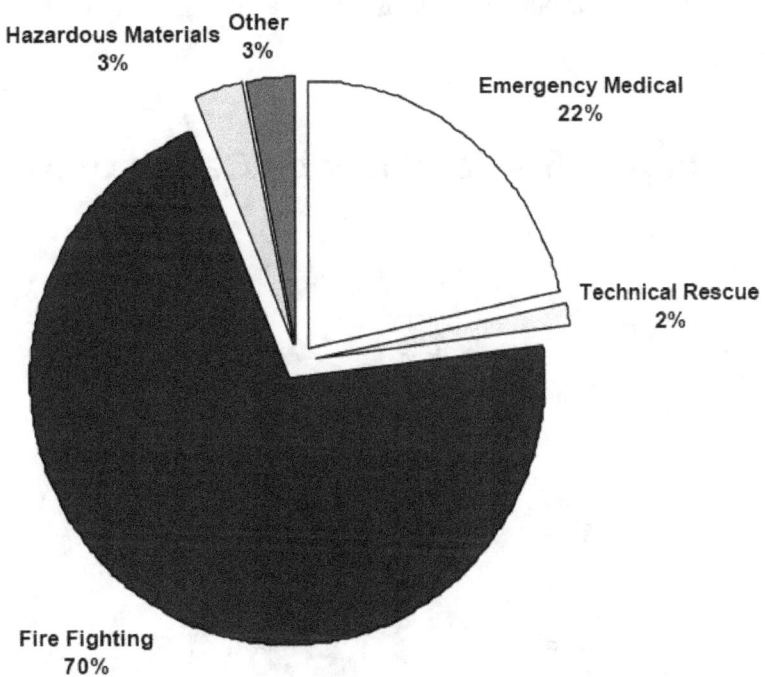

On-Scene or Responding – 65 of 91

Cause of Fatal Injury

As used in this study, the term 'cause of injury' refers to the action, lack of action, or circumstances that directly resulted in the fatal injury, while the term nature of injury refers to the medical cause of the fatal injury or illness, often referred to as the physiological cause of death. A fatal injury usually is the result of a chain of events, the first of which is recorded as the cause. For example, if a firefighter is struck by a collapsing wall, becomes trapped in the debris, runs out of air before being rescued, and dies of asphyxiation, the cause of the fatal injury is recorded as "struck by collapsing wall" and the nature of the fatal injury is "asphyxiation". Similarly, if a wildland firefighter is overrun by a fire and dies of burns, the cause of the death would be listed as "caught/trapped," and the nature of death would be "burns". This follows the convention used in NFIRS casualty reports.

Figure 5 shows the distribution of deaths by cause of fatal injury or illness and Table 2 presents the exact number.

Figure 5 – Fatalities by Cause of Fatal Injury

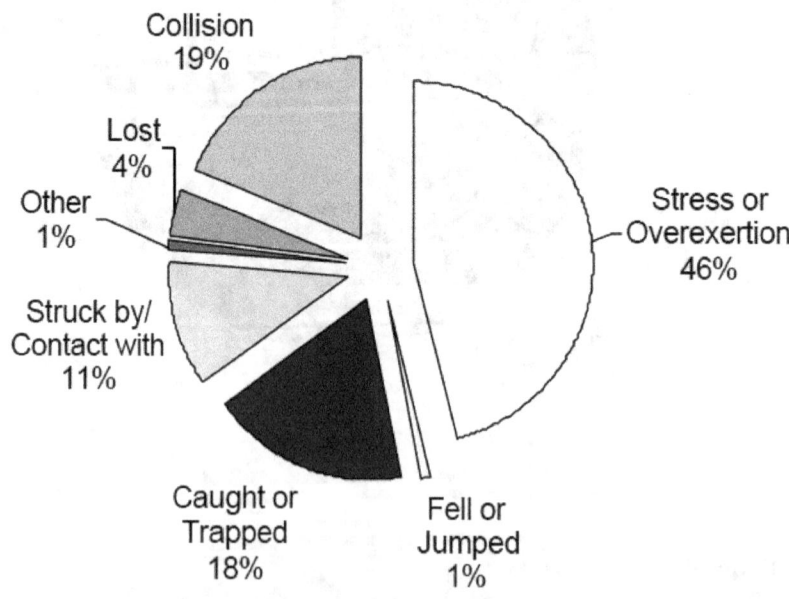

Table 2. Cause of Fatal Injury - 1998	Number	Percent
Stress or Overexertion	42	46%
Collisions	17	19%
Caught or Trapped	16	18%
Struck By/Contact with an Object	10	11%
Lost	4	4%
Fell or Jumped	1	1%
Other	1	1%
	91	100%

As in most previous years, the largest cause category is stress or overexertion, which was listed as the primary factor in 46 percent of the deaths, up from 43 percent in 1997, but returning closer to the 50 percent level that has been experienced in the last several years. Fire fighting is extremely strenuous physical work and is likely one of the most physically demanding activities that the human body performs. Most firefighter deaths attributed to stress result from heart attacks. Of the 42 stress-related fatalities in 1998, 38 firefighters died of heart attacks, two died of CVA's, one died of a gastric hemorrhage, and one died of a seizure. Thirteen of the 42 deaths whose cause is listed as stress/exertion occurred during non-emergency activities.

Seventeen firefighters were killed in collisions. Three were killed in the crash of their fire department helicopter and three were killed in two separate airtanker crashes while fighting wildland fires. The other eleven involved motor vehicle collisions. Two wildland firefighters were killed in separate apparatus collisions in California, one firefighter was killed while on his way to a training class, two firefighters were killed while driving their personal vehicles to emergencies in Oklahoma and Texas, one firefighter who was an off-duty Sheriff's deputy was killed as he responded as a part of his duties as a volunteer firefighter to a trailer fire in his cruiser, and one firefighter was killed when his tractor rolled as he was performing maintenance on a fire road. Four firefighters were killed in rollovers, one involved a pumper enroute to standby duty, one involved a tanker (tender) responding to a fire, one involved a pumper responding to an illegal burn, and one firefighter died when he was ejected from the pumper he was driving back from an EMS incident.

The third leading cause of firefighter fatalities was being caught or trapped, which accounted for 16 deaths (18 percent), up more than six percent from 1997. Five firefighters died after becoming trapped by collapses and eight firefighters were trapped by rapid fire progress in Ohio, Chicago, New York City, and Arkansas. One firefighter drowned while attempting a rescue, one firefighter was trapped by the movement of a fallen tree he was in the process of removing from the roadway, and one firefighter was caught and crushed between a utility pole and a pumper.

The next leading cause of firefighter fatalities for 1998, was being struck by or coming into contact with an object. There were ten deaths in this category including two firefighters in Kansas that were electrocuted when a ladder that they were repositioning came into contact with an electrical line and two firefighters that were struck and killed by pieces of an 18,000 gallon propane tank after it experienced a BLEVE. Five firefighters were killed when they were struck by other motor vehicles. Two were fire police officers directing traffic, one was a firefighter crossing a road at a drill, one firefighter was struck while he worked in the road median, and one firefighter was killed in a chain reaction crash as he worked in the back of an ambulance at an EMS call. One firefighter was struck by a falling parapet wall as he opened doors at a recycling warehouse to allow for a defensive fire attack.

Four firefighters died when they became lost inside of burning structures. One firefighter died in a pet food processing plant in Los Angeles, one firefighter became lost in a supermarket fire in West Virginia, and two firefighters died when they became lost in a auto salvage yard storage facility in North Carolina.

One firefighter was killed due to a fall from a fire department pickup. He was riding in the bed of the pickup while setting up for competition training, he fell and hit his head on the pavement. Another firefighter died when a blood clot became lodged in his lungs.

Nature of Fatal Injury

Table 3 and Figure 6 show the distribution of the 91 deaths by the medical nature of the fatal injury or illness. The leading nature of death in 1998 was heart attacks, which accounted for 38 firefighter fatalities (two more than in 1997).

Table 3. Nature of Fatal Injury	Number	Percent
Heart Attacks	38	42%
Internal Trauma	27	30%
Asphyxiation (includes drowning)	15	17%
CVA/Stroke	3	3%
Burns	3	3%
Electrocution	2	2%
Other	2	2%
Amputation	1	1%
TOTAL	91	100%

Figure 6 – Fatalities by Nature of Fatal Injury

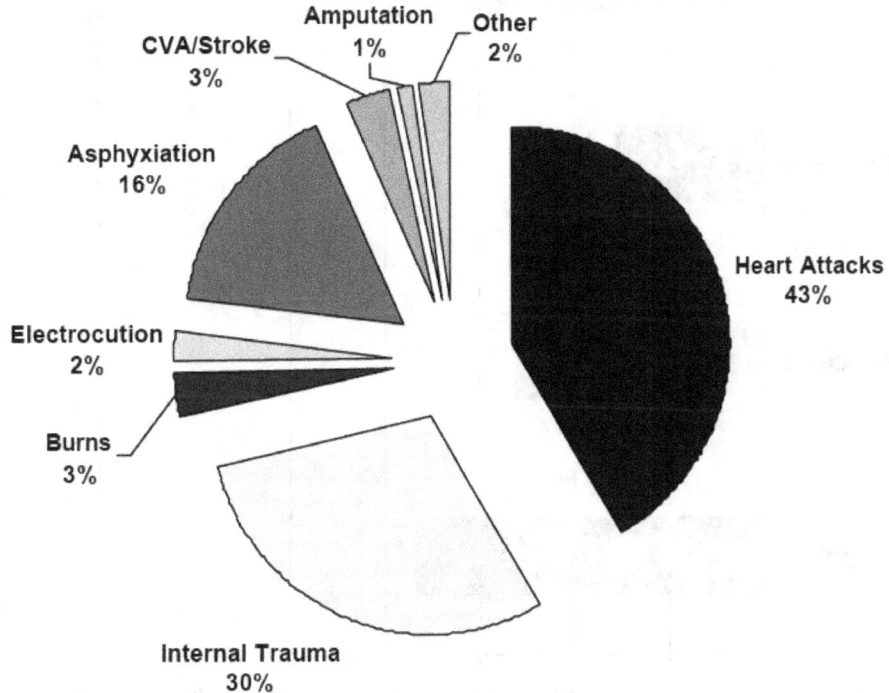

Figure 6a provides a detailed breakdown of heart attacks by type of duty. Seventeen of the heart attacks occurred at the fire scene and six occurred while enroute to or returning from an emergency incident, including one firefighter who died in his personal vehicle as he arrived on the scene of an emergency to perform fire police duties. At the conclusion of the incident, he was found by other firefighters at the wheel of his vehicle, with the vehicle running, in drive, with the firefighter's foot on the brake pedal. Seven occurred at training incidents and five occurred during non-fire incidents. Three heart attacks occurred during other on-duty situations including a heart attack that struck down a firefighter as he slept.

Figure 6a - Heart Attacks by Type of Duty

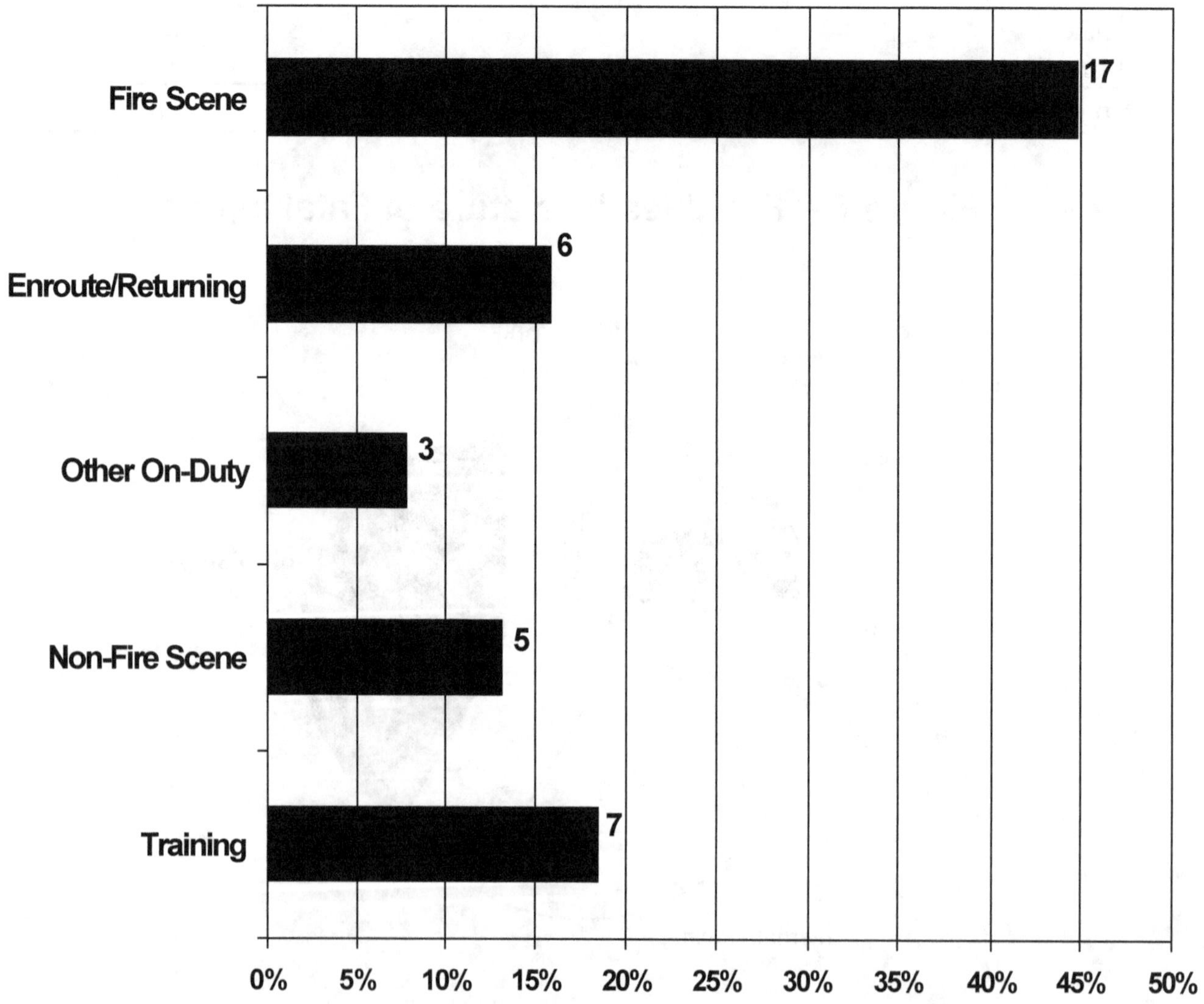

Internal trauma was the second leading nature of death, responsible for 27 deaths (down from 32 in 1997). This total includes 11 firefighters who were involved in vehicle accidents, five firefighters who were hit by vehicles while on the emergency scene or training, three firefighters who died in wildland fire fighting aircraft crashes, three firefighters that died in the crash of a fire department helicopter, two firefighters that were killed in a BLEVE, one death as a result of a fall from a fire department pickup, one firefighter that was crushed by a falling wall, and one firefighter who died in a basement fire.

Asphyxiation was the third leading medical reason for firefighter deaths, responsible for 15 deaths, the same as in 1997, and up from five deaths from 1996. Four incidents claimed nine firefighters. Fourteen of these deaths occurred during structural fire fighting and one was a drowning.

Three of the 91 firefighter fatalities that occurred in 1998 were attributed to burns. One firefighter was severely burned when fire progress overran his bulldozer while he was plowing a fireline, one firefighter was burned when a floor collapse occurred and dropped the firefighter into a fire area, and one firefighter was burned to death when his cruiser collided with a utility pole and burned.

Three firefighters were felled by strokes (CVA's), one of which occurred in the fire station. One stroke occurred after a chief officer complained of not feeling well at a structure fire, he died the next day at home. One stroke/seizure resulted from head injuries that were received in a training maze.

Two firefighters died from electrocution. They were electrocuted while they repositioned a ladder that was going to be loaned to painting contractors working on a church.

One firefighter died when a blood clot became lodged in his lung as he slept, one died of a gastric hemorrhage, and one died of blood loss as the result of a partial leg amputation.

Firefighters Ages

Figure 7 shows the distribution of firefighter deaths by age and cause of death. Younger firefighters were more likely to have died as a result of traumatic injuries from an apparatus accident or after becoming caught or trapped during fire fighting operations. Stress was shown to play an increasing role in firefighter deaths as age increased.

Figure 7 - Fatalities by Age and Cause

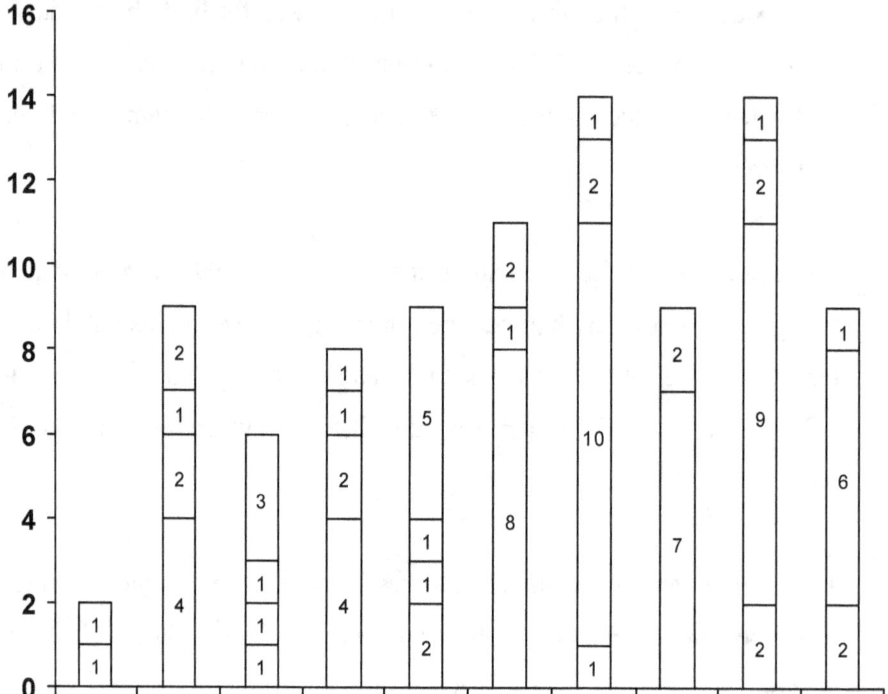

	Under 21	21 to 25	26 to 30	31 to 35	36 to 40	41 to 45	46 to 50	51 to 55	56 to 60	61 and Over
Fall	1									
Caught/Trapped		2	3	1	5	2	1		1	1
Struck by/or contact with object		1	1		1	1	2	2	2	
Other				1						
Stress/Overexertion				2		8	10	7	9	6
Lost		2	1		1					
Collision	1	4	1	4	2		1		2	2

This is also reflected in Figure 8, which shows the distribution of deaths by age and medical nature of injury. Trauma and asphyxiation were responsible for most of the deaths of younger firefighters, while heart attacks were much more prevalent among older firefighters. Heart attacks accounted for 65 percent of firefighter deaths where the firefighter was age 41 or higher.

Figure 8 – Fatalities by Age and Nature

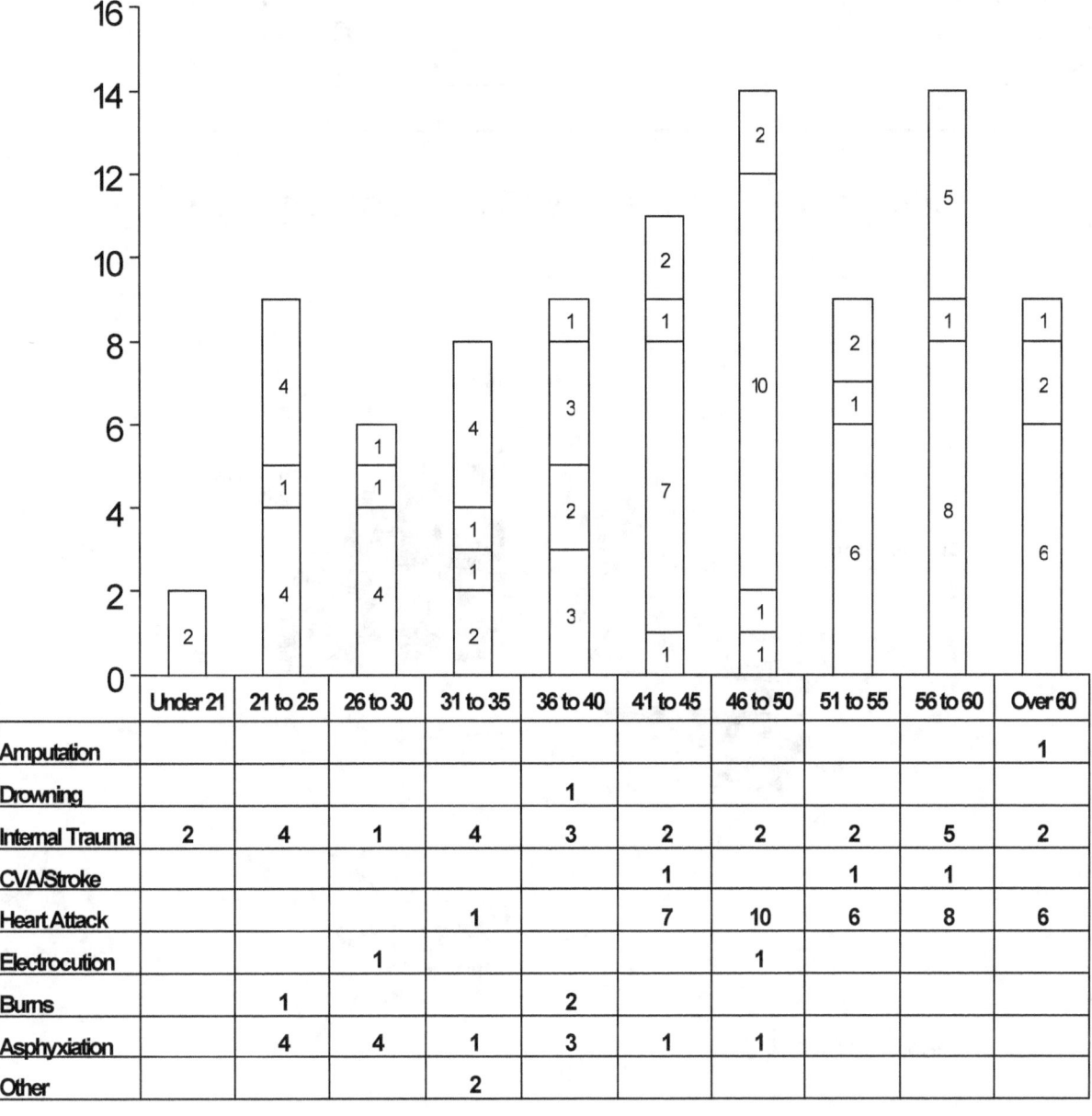

	Under 21	21 to 25	26 to 30	31 to 35	36 to 40	41 to 45	46 to 50	51 to 55	56 to 60	Over 60
Amputation										1
Drowning					1					
Internal Trauma	2	4	1	4	3	2	2	2	5	2
CVA/Stroke						1		1	1	
Heart Attack				1		7	10	6	8	6
Electrocution			1				1			
Burns		1			2					
Asphyxiation		4	4	1	3	1	1			
Other				2						

Fixed Property Type

There were 42 fireground deaths in 1998. Figure 9 and Table 4 show the distribution by fixed property use.

Table 4. Property Use for Fireground Deaths	Number	Percent
Residential	17	40%
Commercial	11	26%
Outdoor Property	7	17%
Street / Road	3	7%
Manufacturing	2	5%
Storage	2	5%
TOTAL	42	100%

Figure 9 – Fatalities by Fixed Property Use

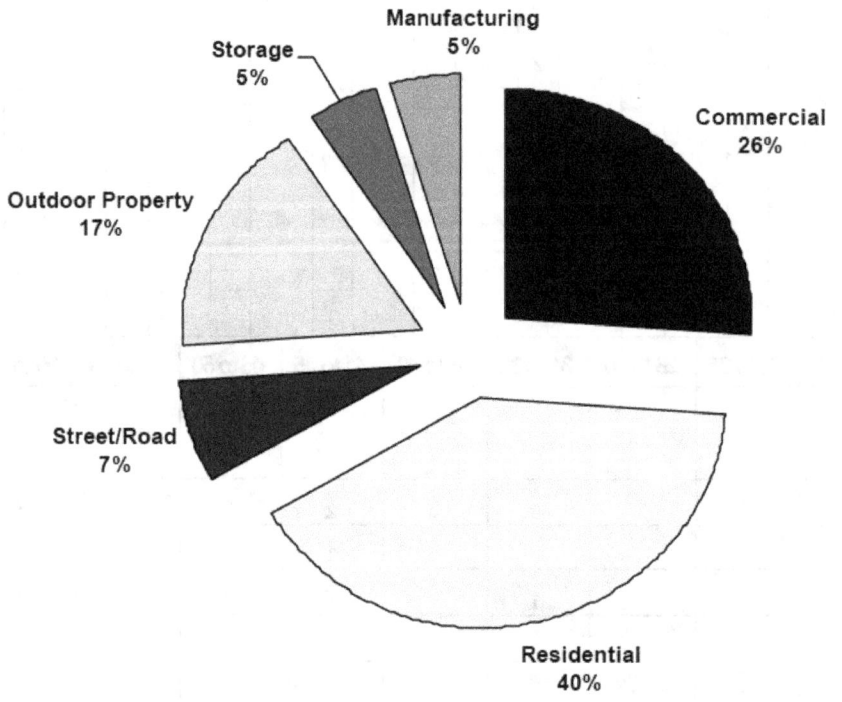

Fireground deaths only - 42 of 91

Fireground Deaths

Structural fires accounted for 32 fireground deaths. As in most years, residential occupancies accounted for the highest number of these fireground fatalities, with 17 deaths (half of all structural fire deaths). Residential occupancies usually account for 70-80 percent of all structure fires and a similar percentage of the civilian fire deaths each year[2]. The frequency of firefighter deaths in relation to the number of fires is much higher for non-residential structures.

Fires that occurred on outdoor properties and "street/road" accounted for a total of ten deaths. Three deaths were the result of wildland fire fighting aircraft crashes, three were heart attacks at the scene of vehicle fires, one firefighter died of burns when his bulldozer was overrun by fire progress, one firefighter died of a heart attack while filling brush trucks with water, a firefighter died of a heart attack after working a wildland fire for four hours, and one firefighter was crushed between his pumper and a utility pole as the truck was relocated at a wildland fire.

Type of Activity

Figure 10 and Table 5 show the type of fireground activity that the 42 firefighters were engaged in at the time they sustained their fatal injuries or illnesses.

Table 5. Type of Activity for Fireground Deaths	Number	Percent
Advancing Hose Lines / Fire Attack	18	42%
Search and Rescue	7	17%
Water Supply	4	10%
Support Duties	3	7%
Ventilation	3	7%
Other	3	7%
Incident Command	2	5%
Cutting Fire Breaks (Wildland)	2	5%
TOTAL	42	100%

---

[2] Complete NFIRS data for 1998 fire incidence was not available at the time of this report, but residential fires typically account for between 70 and 80 percent of all civilian fatalities each year.

Figure 10 – Fatalities by Type of Activity

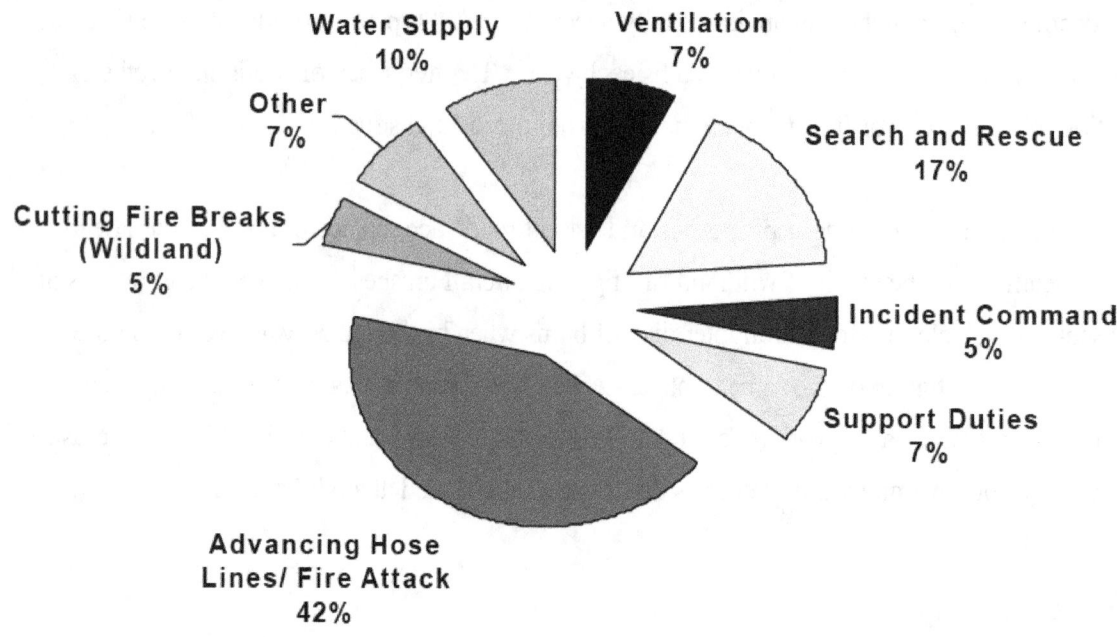

Fireground Operations Deaths Only - 42 of 91

Eighteen firefighters died while engaged in fire attack and advancing hose lines. This is a small but significant decrease compared to the 21 firefighters killed performing this activity in 1997. Nine of the eighteen firefighters were killed engaged in structural fire fighting, two were killed in explosions, three were killed in wildland fire fighting aircraft crashes, two others were killed at wildland fires, one firefighter died at a dumpster fire, and one died at a vehicle fire.

Seven firefighters were killed during search and rescue operations (up two from the levels experienced in 1997 and 1996). Two were trapped in a collapse of a residential structure fire as they searched for a fire victim, three were overcome by rapid fire progress while searching a high rise apartment building, one firefighter was killed searching for fire in a supermarket, and one firefighter was killed as he attempted to rescue another firefighter in an auto salvage yard storage building.

Four firefighters were killed while engaged in water supply operations (the same level as 1997). Two died at structural fires and two at wildland fires.

Three firefighters died while performing support duties at fires, all three were structural fires. Three firefighters were killed while performing ventilation duties, one fell through the roof at a structural fire, one had entered a residential structure by ground ladder when he succumbed to a heart attack, and one was prying open the hood of a car when he experienced a heart attack.

Two firefighters died while cutting fire breaks, one when his bulldozer was overrun by fire and one of a heart attack. Two firefighters died while performing incident command tasks, one at a car fire and the other at a structural fire.

Time of Alarm

The distribution of all 1998 firefighter deaths according to the time of day when the fatal injury occurred is illustrated in Figure 11 (46 times were not reported). For structural fire fighting deaths only, 71% occurred between 8:00 p.m. and 8:00 a.m.

Figure 11 - Fatalties by Time of Fatal Injury (1998)

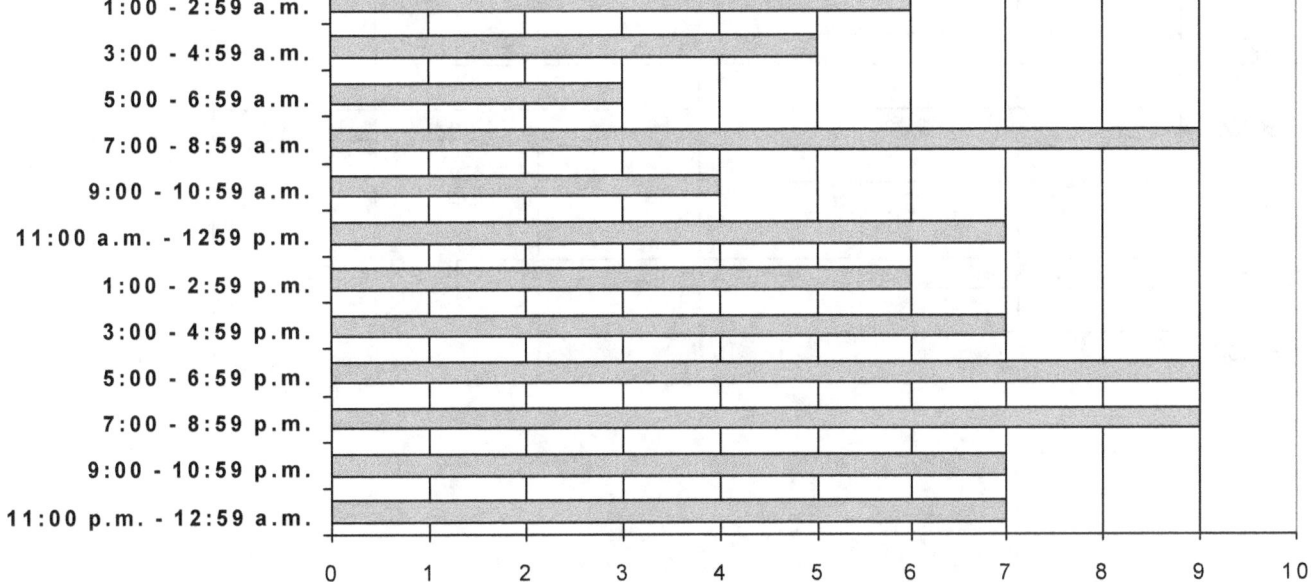

12 times were not reported

Month of the Year

Figure 12 illustrates firefighter fatalities by month of the year. Firefighter fatalities peaked in September. Other than the fact that wildland fires occur in the wildland season, no trends were identified.

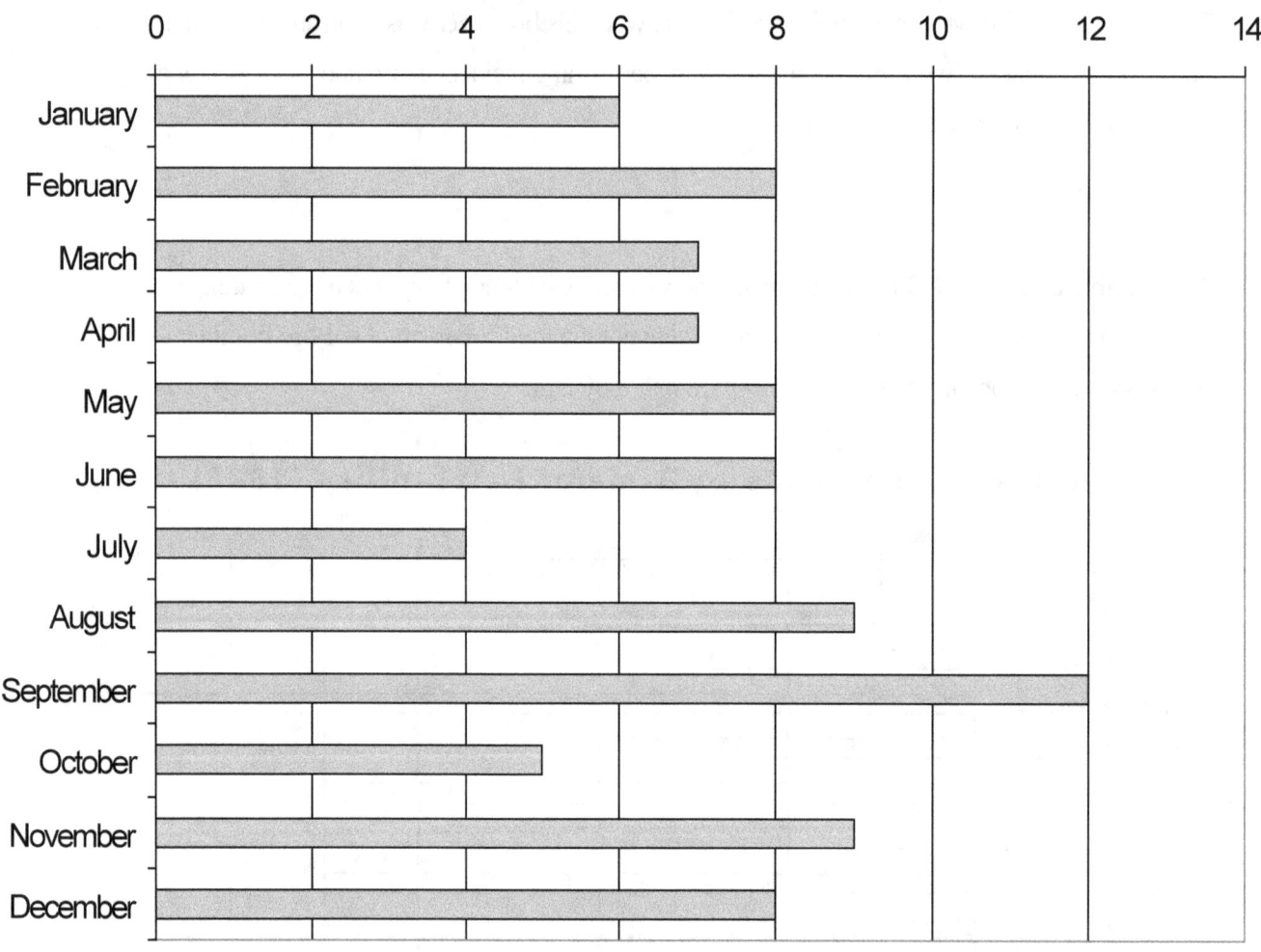

Figure 12 - Deaths by Month of the Year

State and Region

The distribution of firefighter deaths by state is shown in Table 6.[3] Thirty-three states had at least one firefighter fatality. New York had the highest number of deaths with 15 followed by California with 9. Figure 13 shows the firefighter fatalities divided by region of the country and their status as career, volunteer, or wildland firefighters.

Figure 13.
Firefighter Deaths By Region 1998

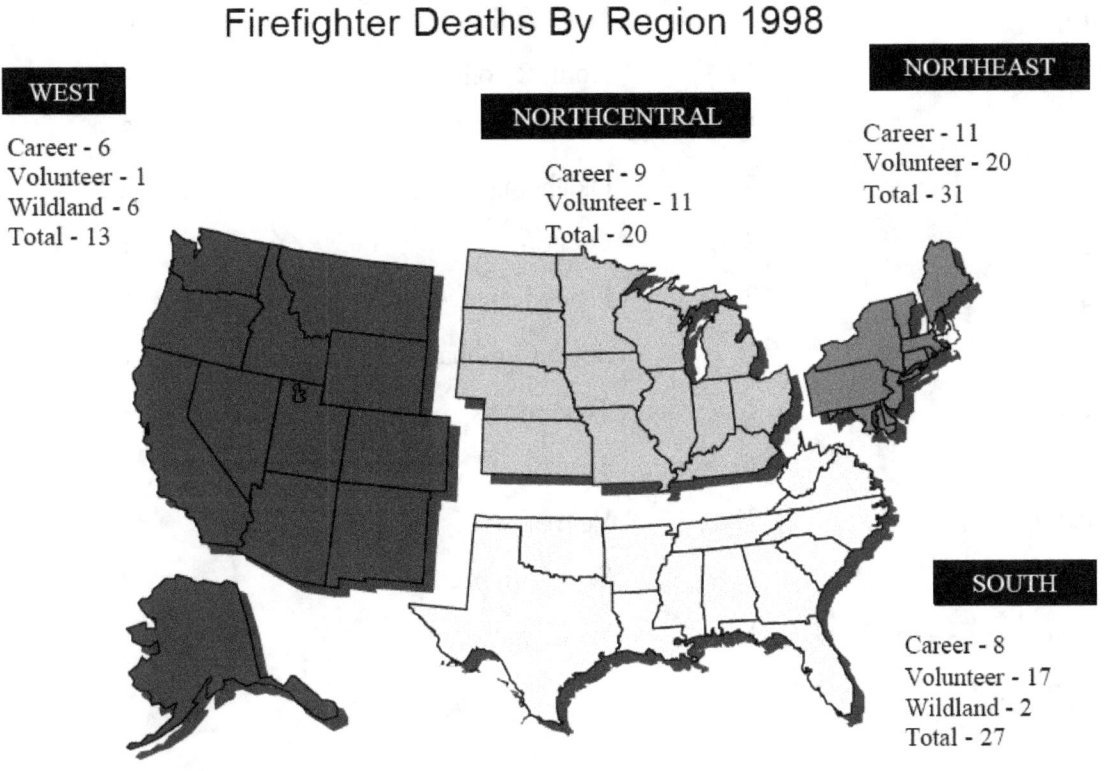

WEST
Career - 6
Volunteer - 1
Wildland - 6
Total - 13

NORTHCENTRAL
Career - 9
Volunteer - 11
Total - 20

NORTHEAST
Career - 11
Volunteer - 20
Total - 31

SOUTH
Career - 8
Volunteer - 17
Wildland - 2
Total - 27

---

[3] This list attributes the deaths according to the state where the fire department or unit is based, as opposed to the state where the death occurred. They are listed by those states for statistical purposes, and for the National Fallen Firefighter Memorial at the National Emergency Training Center.

Table 6.

1998 State with On-Duty Firefighter Fatalities

State	Number of Deaths	State	Number of Deaths
Alabama	2	Missouri	1
Arkansas	1	Montana	2
California	9	New Jersey	1
Delaware	1	New Mexico	1
Florida	1	New York	15
Georgia	1	North Carolina	4
Illinois	5	Ohio	3
Indiana	5	Oklahoma	2
Iowa	2	Oregon	1
Kansas	2	Pennsylvania	6
Kentucky	1	South Carolina	1
Louisiana	1	Tennessee	3
Maine	1	Texas	6
Maryland	4	Vermont	2
Massachusetts	1	West Virginia	2
Minnesota	1	Wisconsin	1
Mississippi	2		

Total - 91

Analysis of Urban/Rural/Suburban Patterns in Firefighter Fatalities

The US Bureau of the Census defines "urban" as a place having a population of at least 2,500 or lying within a designated urban area. Rural is defined as any community that is not urban. Suburban is not a census term but may be taken to refer to any place, urban or rural, that lies within a metropolitan area defined by the Census Bureau, but not within one of the central cities of that metropolitan area.

Fire department areas of responsibility do not always conform to the boundaries used for the census. For example, fire departments organized by counties or special fire protection districts may have both urban and rural coverage areas. In such cases, it may not be possible to characterize the entire coverage area of the fire department as rural or urban, and firefighter deaths were listed as urban or rural based on the particular community or location in which the fatality occurred.

The following patterns were found for 1998 firefighter fatalities. These statistics are based on answers from the fire departments and where no data from the department was available, the data is based upon population and area served reported by the fire departments.

Table 7.

	Urban/Suburban	Rural	Federal or State Parks/Wildland	Total
Firefighter Deaths	46	38	7	91

SPECIAL TOPICS

Back to Basics

In 1998, firefighters continued to die at least partly as the result of violations of basic safety rules. The use of seat belts, the use of Self Contained Breathing Apparatus (SCBA), limitations on where firefighters can ride on moving fire apparatus, and the use of PASS devices are basic safety concepts that can save firefighter's lives. Nothing about these areas is new, all four are specifically addressed in NFPA 1500, ***Standard on Fire Department Occupational Safety and Health Program***. The only way that these concepts will work is through the development of necessary procedures, provision and proper maintenance of needed equipment, and – most importantly - constant vigilance on the part of every individual firefighter and fire officer.

Seat Belts

At least six of the eleven firefighters killed in vehicle collisions were not wearing seat belts. The lives of Firefighter Timothy Allen, Training Officer Brian Cannon, Firefighter Allen Heirtzler, Junior Firefighter Jake Hoeffner, Firefighter Patrick McKinney, and Assistant Chief Richard Rice might have been spared if they had taken the simple precaution of fastening their seat belts. Most of these collisions were survivable – three of these incidents involved the survival of passengers in the firefighter's vehicle. Four of these firefighters were partially or fully ejected from their vehicles.

Firefighter's lives are at risk during every emergency response. A study of fire apparatus traffic collisions was performed for the Freightliner Corporation by the University of Michigan Transportation Research Institute in 1998. The study looked at fire apparatus collision experience for 1994-1996. The study found the following:

> ***There are 2,472 Fire Apparatus Accidents Each Year***
> ***Six Occupants of Fire Apparatus are Killed Each Year***
> ***413 Occupants of Fire Apparatus are Injured Each Year***
> ***20% of Fire Apparatus Collisions Result in Rollovers***
> ***47% of Fire Apparatus Collisions are at Intersections***
> ***Almost all Firefighter Fatalities Were From Rollovers – 73%***
> ***76% of the Firefighters Killed Were Not Wearing Seat Belts***

In addition to the firefighter injuries and deaths, the study found that 21 civilians are killed each year as the result of collisions with fire apparatus and 642 are injured.

SCBA Use

One firefighter killed in 1998 was not wearing an SCBA when he suffered a heart attack while fighting a structural fire. At the time Lieutenant Stephen Murphy was stricken with the heart attack, he was inside of the involved structure conducting ventilation activities. While there may not be any direct connection between the lack of SCBA use and this firefighter's death, this serves as a reminder of the need to wear and use an SCBA. NFPA 1500, Standard on Fire Department Occupational Safety and Health Program, requires SCBA use when firefighters are engaged in any operations where they might encounter atmospheres that are known to be Immediately Dangerous to Life and Health (IDLH), atmospheres that are potentially IDLH, or where the safety of the atmosphere is unknown.

Vehicle Riding Positions

Two firefighters were killed in 1998 as they rode in exposed positions on fire department vehicles. Firefighter Barvon Hamilton was killed as he rode the back step of a pumper and Junior Firefighter Jake Hoeffner was killed when he fell from the bed of a fire department pickup. In addition to riding in an unsafe location, neither firefighter was wearing any form of restraint device.

Riding in exposed positions on fire department vehicles cannot be allowed under any circumstances. Firefighter Hamilton was riding the back step for a short trip to another fire location and Firefighter Hoeffner was riding in the bed of a pickup as it crossed a parking lot.

National Fire Protection Association Standards call for the provision of an interior seated and belted position for each firefighter expected to ride a piece of fire apparatus since 1991. Apparatus manufactured prior to the implementation of the standard will be in use for decades to come. There are a number of alternatives to riding the back step of apparatus including cab modifications to existing apparatus, the use of smaller vehicles to transport firefighters that cannot safely ride on apparatus, and construction of safe riding compartments on existing apparatus.

Each fire department should have a policy in place that makes riding in exposed positions on fire department vehicles, even for short trips, unacceptable.

Personal Alert Safety Systems (PASS Devices)

PASS devices are small electronic boxes that attach to a turnout coat or SCBA. They are about the size of a pack of cigarettes. They are designed to sound a loud alarm if a firefighter is motionless for more than thirty seconds. The alarm is intended to guide other firefighters to the firefighter in need of assistance. PASS devices may be stand-alone or they may be integrated into the SCBA and activate automatically when the SCBA is activated (turned on).

The PASS status of many firefighters who died in 1998 is unknown. At least two firefighters who died in 1998 did so while wearing PASS devices that were not turned on or armed. First Lieutenant Robby Blizzard and Firefighter Gregory Carter were reported to be wearing PASS devices but they did not activate them. Both firefighters most certainly knew that they were in trouble, Firefighter Blizzard was attempting to rescue the Fire Chief of a neighboring fire department and Firefighter Carter was out of air and placed his breathing air tube inside of his coat, yet neither firefighter remembered and activated his PASS.

Many firefighters neglect to turn on or activate their stand-alone PASS devices as they enter the emergency scene. Reasons that are cited for this lack of action include forgetfulness, being occupied with the task at hand, and a desire to avoid PASS false alarms. Even when confronted with emergencies, such as being lost or trapped in a structure, firefighters seldom remember to activate their PASS devices.

The accuracy and sensitivity of PASS devices has improved dramatically since they were introduced in the mid-1980's. The latest edition of the NFPA standard on PASS devices, NFPA 1982 - *Standard on Personal Alert Safety Systems (PASS) for Fire Fighters*, requires automatic activation for stand-alone and integrated PASS devices.

Most SCBA manufacturers offer PASS devices that are integrated into the SCBA. When the SCBA is turned on or activated, the PASS device is armed. The PASS cannot be deactivated until the SCBA is shut off and drained of air. These PASS devices are also available in versions that can be retrofitted into existing SCBA's.

Firefighter Health

In 1998, 41 firefighters died of heart attacks and strokes. In the last four years, 171 firefighters have died of heart attacks and strokes. Many of the firefighters who died had pre-existing medical conditions that placed them at higher risk for heart disease and strokes. In many cases, these conditions were known to the fire department. In many more cases, the health of the firefighter was unknown prior to their death.

In 1998, at least 19 of the 41 firefighters who died of heart attacks had pre-existing conditions such as prior heart problems, bypass surgery, or diabetes which can indicate the possibility of heart and cardiovascular disease.

In order to reduce the terrible toll taken by heart disease and strokes, the following steps can be taken:

> **Firefighters should be required to submit to periodic health assessments to determine their fitness for duty.** NFPA 1982, *Standard on Medical Requirements for Fire Fighters*, provides criteria for use by physicians as they perform physical examinations for firefighters. The standard requires medical evaluations for firefighters prior to their entry into the fire service, when a firefighter returns to duty after an illness, and on an annual basis. The standard describes a limited number of medical conditions that preclude a firefighter from participating in training or fire fighting activities and a larger list of medical conditions that may preclude a firefighter from such activities.

> Of the 41 firefighters who died of heart attacks or strokes in 1998, only eight were required to have annual health checks and seventeen had either never had a medical evaluation or had not had one since they joined their department.

> **Fire Departments should provide their members with information on healthy eating and diets that reduce the risk of heart disease.** Firefighters are like many members of the general public in that they need more information on which foods have been shown to lessen the occurrence of heart disease. Career firefighters often eat together every shift and a change in eating habits for the entire crew may be more acceptable than modification of these habits alone. Free information is available from a number of sources including the American Heart Association at **http://www.american heart.org** and from the American Diabetes Association at **http://www.diabetes.org**

Firefighters should participate in physical fitness programs. The physical requirements of fire fighting are well known. A physical fitness program sponsored by the fire department may improve individual health, individual effectiveness, and may also be seen in more efficient fire fighting activity. There are a number of options available for such programs, some involve the purchase of fitness equipment for fire stations and others enlist the help of local fitness clubs, school fitness facilities, and local parks.

Fire Departments should assure that immediate medical care is available at every emergency scene. NFPA 1500 has long required the availability of basic level life support at emergency scenes and higher level life support at special operations incidents. Firefighters should be trained in basic skills that can be used to save the people we serve **and ourselves**. Every firefighter should be trained in at least basic first aid and CPR.

Fire Departments should ensure that an Automatic External Defibrillator (AED) or paramedic level care is available on every incident scene. In the past decade, the cost and level of training needed to operate an AED have dropped considerably. AED's can be found in malls, airports, and state legislatures across the country. While they are not a substitute for paramedic level Advanced Life Support (ALS), they do save lives. With the number of firefighters who die each year from heart attacks, the provision of this life saving tool at every incident scene has the potential to save a number of lives each year.

CONCLUSIONS

The analysis of firefighter deaths in 1998 indicates that the overall long-term trend toward fewer firefighter fatalities is continuing. The 91 fatalities in 1998 are the third lowest recorded since the inception of this study, and only the sixth time the total number of fatalities has dropped below 100. Four of the last five years have resulted in less than 100 firefighter fatalities. Firefighters die while engaged in a wide range of services to the community. The addition of emergency medical services provision, hazardous materials response, and technical rescue service has brought with them risks and accompanying firefighter deaths.

Although the long term reduction in the number of deaths is reason for hope, firefighters are still dying at an unacceptable rate.

Stress-induced heart attacks remained the top cause of firefighter deaths. Continued focus on firefighter health and wellness may likely reduce the impact of this killer in the future. There have been a number of studies of this issue including the recent ***Fire Service Joint Labor Management Wellness-Fitness Initiative*** which was developed cooperatively by the International Association of Fire Chiefs (IAFC), the International Association of Fire Fighters (IAFF), and members of ten fire departments. This document, released in 1997, provides a framework for long term attention to issues of firefighter health and wellness. Copies of this report are available for sale from the IAFC. Copies are also available to IAFF members from the IAFF. A cost may be involved.

International Association of Fire Chiefs
4025 Fair Ridge Drive
Fairfax, VA 22033-2868
http://www.iafc.org

International Association of Fire Fighters
1750 New York Avenue, NW
Washington, DC 20006
http://www.iaff.org

Internal trauma was the second leading nature of death for 1998. This total includes eleven firefighters who were involved in vehicle collisions and five firefighters who were struck and killed by vehicles at the emergency scene. Three deaths involving flight crews of wildland fire fighting aircraft remind us of the hazards associated with their deployment.

Of the eleven firefighters killed in vehicle collisions, at least six were not wearing seat belts, a basic safety precaution.

Scene safety was addressed in the 1997 report on firefighter fatalities and has remained a concern. Three of the firefighters killed in 1998 when they were struck by other motor vehicles were members of the Fire Police. The other two firefighters killed when they were struck at an emergency scene were involved in emergency medical calls and were off of the roadway when they were struck.

One firefighter death occurred in 1998 as the result of an injury that occurred in a previous year. Two Puerto Rican firefighters died while performing their duties. While their deaths are not included in this analysis, these firefighters will be included in the 1998 annual Fallen Firefighter Memorial Service at the National Emergency Training Center, and their names will be included on the list of firefighters who died in 1998.

Fourteen firefighters died of asphyxiation in structure fires. An operating PASS device may not have saved all of them but would have likely changed the outcome of some of these incidents. In at least two cases, PASS devices were worn but not activated. Most SCBA manufacturers offer integrated PASS devices that activate when the SCBA is utilized. Kits are available from most manufacturers to retrofit existing SCBA's with this life saving device.

New developments in respiratory protection standards and their impact on fire department operations, and therefore firefighter safety, will begin to be seen in the next few years.

APPENDIX A
SUMMARY OF 1998 INCIDENTS

If additional information is available regarding a firefighter fatality, the reader is directed to these sources. Contact information for these sources is provided at the end of the appendix.

1/5/98
Harold E. Roemer, Jr., Firefighter
Age 55, Volunteer
Greenlawn Fire District, New York

Firefighter Roemer had just completed thirty minutes on a treadmill in the gym located at Fire Department Headquarters. He signed out of the gym, took a drink of water, and collapsed due to a heart attack.

1/6/98
Prince Albert Mousley, Jr., Firefighter
Age 58, Career
Wilmington Fire Department, Delaware

Firefighter Mousley was a member of a ladder company on the scene of an oil burner fire in the basement of a residential structure. As he and a partner entered the rear of the structure, Firefighter Mousley stated that he was tired and collapsed of a heart attack. Firefighter Mousley had just returned to duty after a battle with cancer. Further information related to this incident can be found in NIOSH Fire Fighter Fatality Investigation 98-F-13.

1/12/98
Robert J. O'Toole, Firefighter
Age 25, Part-Time Paid
Washington Township Fire Department, Ohio

Firefighter O'Toole responded to an automobile collision on an interstate highway. The victim of the original collision had been loaded into an ambulance and had left the scene. As Firefighter O'Toole and others began to disconnect the battery on the vehicle which was located in the median, he was struck and killed by another vehicle that had lost control on the ice. A police officer was also killed and another firefighter was severely injured in this incident.

1/16/98
Brian Allen Cannon, Training Officer
Age 30, Volunteer
Taylors Bridge Fire Department, Inc., North Carolina

Firefighter Cannon and a Fire Captain were returning to the station from the scene of a traffic collision. The pumper left the roadway and overturned. Firefighter Cannon, who was not wearing a seat belt, was ejected and sustained blunt trauma injuries to the head.

1/21/98
Gregory Scott Carter, Firefighter
Age 24, Volunteer
Fairlea Volunteer Fire Department, West Virginia

Firefighter Carter responded to a report of smoke in a supermarket. The market was contained in a strip mall which also included a post office and a photo-processing store. Firefighter Carter had been employed at the supermarket in the past. Firefighter Carter and a Captain entered the front of the store in full protective clothing and SCBA to search for the fire. They became disoriented while trying to exit the store. The Captain alerted other firefighters by radio that he and Firefighter Carter were lost and in need of rescue. Firefighter Carter ran out of air and placed the breathing tube from his SCBA into his coat in an attempt to breathe. The Captain was able to escape without significant injury. Immediate attempts were made by on scene firefighters to rescue Firefighter Carter but rescuers were driven back by intense heat and smoke. Firefighter Carter was wearing a PASS device but it was not turned on. No hose line or search rope was used. The cause of death was smoke and soot inhalation, carbon monoxide poisoning, and complete body charring. This was an accidental fire caused by an electrical malfunction in a wall. Further information related to this incident can be found in NIOSH Fire Fighter Fatality Investigation 98-F-04.

1/27/98
Stephen Earl Murphy, Lieutenant
Age 47, Career
Philadelphia Fire Department, Pennsylvania

Lieutenant Murphy responded to a fire in a rowhouse dwelling. He carried a sixteen-foot portable ladder to the rear of the structure, raised the ladder, broke out windows, climbed the ladder, and entered a bedroom. Other firefighters who ascended the ladder reported seeing Lieutenant Murphy in the bedroom. He ordered them to proceed into the structure and continue ventilation. When the firefighters returned to the bedroom, they found Lieutenant Murphy face down and unresponsive suffering from an apparent heart attack. Emergency medical treatment was initiated and Lieutenant Murphy was transported to the hospital where a heartbeat was restored. Lieutenant Murphy died on February 3, 1998. The cause of the fire was ruled accidental as a result of a portable kerosene heater placed too close to combustibles. Lieutenant Murphy was not wearing an SCBA.

2/5/98
Stephen D. Carletti, Firefighter, Age 43
David P. Theisen, Firefighter, Age 29
Volunteer
Crooksville Fire Department, Ohio

Firefighter Carletti and Firefighter Theisen responded to a report of a fire in the basement of a single story home. They entered the basement with other firefighters and extinguished fire in the ceiling. In the process of moving around the basement, the attack line was pinched off when it was caught in a folding chair. Firefighters were not aware that their water supply had been cut off. When they began to pull additional ceiling tiles, the room experienced a flashover. Of the five firefighters in the basement when the flashover occurred, two escaped, one was rescued, and two were killed. An adjacent room, which had not been discovered by the firefighters, was fully involved in fire and fire spread to the other room when tiles were removed. Repeated radio requests for help and water were received from the basement but rescuers were unable to reach the firefighters in distress due to severe heat and fire. Both firefighters were wearing their PASS devices, they were turned on, and they activated. The fire cause was determined to be accidental. Firefighter Carletti died of asphyxiation and burns and Firefighter Theisen died as the result of a crushing injury. Firefighter Theisen was also a career firefighter in Westerville. The Crooksville Fire Department suffered a firefighter fatality in 1997. Further information related to this incident can be found in NIOSH Fire Fighter Fatality Investigation 98-F-06.

2/10/98
Richard L. Kalous, Firefighter
Age 50, Career
De Pere Fire Rescue, Wisconsin

Firefighter Kalous responded as a member of an engine company to a car fire. Upon arrival at the scene, he hand stretched a five-inch supply line to a fire hydrant about 75 feet from the engine. When it was determined that the supply line would not be needed, he was directed to don an SCBA and assist with fire attack. He was discovered by other firefighters on a side step of the engine unresponsive and suffering from a heart attack. Despite the immediate efforts of firefighters at the scene and other responding firefighters, he was not successfully revived.

2/11/98
Warren D. Myers, Jr., Firefighter
Age 48, Career
Tulsa Fire Department, Oklahoma

Firefighter Myers and his crew responded to a gas leak at a single-family dwelling. The line was shut off and, after repairs were made, Firefighter Myers turned the gas back on. Firefighter Myers was observed to be fatigued at the incident. When his engine company returned to quarters, Firefighter Myers did not get off the truck and was seen to be in distress. Despite immediate medical treatment by his crew and others, Firefighter Myers died of a heart attack. Further information related to this incident can be found in NIOSH Fire Fighter Fatality Investigation 98-F-29.

2/11/98
Patrick Joseph King, Firefighter Paramedic, Age 40
Anthony E. Lockhart, Firefighter, Age 40
Career
Chicago Fire Department, Illinois

Firefighter King and Firefighter Lockhart responded on different companies to a report of a structural fire in a tire shop. No visible fire was encountered, there was no excessive heat, and only light smoke was found in most of the building with heavier smoke in the shop area. Ten firefighters were in the interior of the structure when an event that has been described as a flashover or backdraft occurred. The firefighters were disoriented by the effects of the backdraft. Some were able to escape but Firefighter King and Firefighter Lockhart were trapped in the structure. A garage door that self-operated due to fire exposure may have introduced oxygen into the fire area and may have been a factor in the backdraft. The exit efforts of firefighters were complicated by congestion in the building. Within minutes of the backdraft, the building was completely involved in fire and rescue efforts were impossible. Both firefighters died from carbon monoxide poisoning due to inhalation of smoke and soot. Further information related to this incident can be found in NIOSH Fire Fighter Fatality Investigation 98-F-05.

2/17/98
Keith C. Thomas, Fire Chief
Age 56, Volunteer
Aubbeenaubbee Volunteer Fire Department, Indiana

Chief Thomas was helping to prepare for a CPR training session in his fire station when he collapsed and died due to a heart attack.

2/25/98

William E. Bonnar, Sr., Battalion Chief

Age 61, Career

Orland Fire Protection District, Illinois

Chief Bonnar collapsed and died of a heart attack approximately 20-30 minutes after the completion of an SCBA drill in a commercial structure.

3/8/98

Joseph C. Dupee, Fire Captain I

Age 38, Career

Los Angeles City Fire Department, California

Captain Dupee and his company were dispatched to a structure fire in a pet food processing company and were assigned to backup interior crews. When fire conditions worsened, all firefighters exited the building with the exception of Captain Dupee who had somehow been separated from his crew. The situation was further complicated by the activation of an emergency signal by another firefighter that had become disoriented (he was rescued by his company officer). Shortly after firefighters left the building, a partial roof collapse occurred. When it was determined that Captain Dupee was missing, a rapid intervention crew forced entry in the rear of the structure and removed Captain Dupee. He was burned over 95% of his body, was provided with advanced life support care, and pronounced dead at the hospital. The cause of death was determined to be asphyxiation and burns. The fire was accidental and started as a grease fire in a convection oven. Further information related to this incident can be found in NIOSH Fire Fighter Fatality Investigation 98-F-07.

3/8/98

Edward J. Matter, Jr., Firefighter

Age 56, Volunteer

Westford Volunteer Fire Department, New York

Firefighter Matter and other members of his Department were dispatched to a report of a tree down blocking the roadway. Firefighter Matter and another fire department member arrived at the scene in their personal vehicles, each carried his own chain saw. They set to work removing parts of the tree from the roadway. The firefighters agreed that they would use a rope to pull the remnants of the tree to the ground to make a safer operation. While the rope was being prepared, Firefighter Matter continued to remove loose debris and began to use his chain saw to cut at one of the larger branches supporting the tree. Firefighter Matter was caught and carried by the tree as it rotated and became pinned face down under the largest section of the tree. Despite immediate removal of the tree by other firefighters, Firefighter Matter died from crushing injuries.

3/9/98

David John Good, Firefighter

Age 36, Volunteer

Lionville Fire Company, Pennsylvania

Firefighter Good was killed when he was struck by a tractor trailer truck that had lost control and slid into firefighters providing treatment at the scene of an earlier automobile collision. Firefighter Good was in the rear of the ambulance when he was struck. Nine other responders were injured, three of them severely. All emergency response personnel were out of the travel lanes when the incident occurred. The incident occurred in heavy rain.

3/23/98

Michael A. Butler, Firefighter/Lead Paramedic, Age 33

Michael D. McComb, Apparatus Operator, Age 48

Eric F. Reiner, Firefighter/Lead Paramedic, Age 33

Career

Los Angeles City Fire Department, California

These three firefighters died when the fire department helicopter in which they were flying crashed in a park. The crash occurred while they were transporting an eleven year old child to the hospital that had been injured in a vehicle collision. In addition to these fatalities, the child was killed in the crash and the pilot and one additional crew member were severely injured. The reason for the crash has been attributed to the inflight failure of the tail rotor system. More information related to this incident can be found in National Transportation Safety Board report LAX98GA127.

3/28/98

Richard K. Rice, Assistant Chief

Age 38, Volunteer

Nassauville Volunteer Fire Department, Florida

Chief Rice was killed in a vehicle collision while enroute to the scene of an illegal burn. Chief Rice was operating a pumper when the truck left the road, rolled one and a half times, and ended up on its roof in a ditch. Chief Rice was partially ejected and was pronounced dead at the scene.

4/1/98

Jeffrey William Reick, Safety Officer

Age 34, Volunteer

Aurora Fire Department, Indiana

Safety Officer Reick experienced a gastric hemorrhage while setting up for a live fire training exercise. Firefighter Reick was stricken when he exited the structure after another firefighter set the training fires. CPR was initiated immediately but Safety Officer Reick succumbed to this illness on 4/2/98.

4/9/98

Thomas James Archer, Jr., Firefighter, Age 46

Larry R. Walsh, Firefighter, Age 45

Volunteer

Albert City Community Fire Department, Iowa

Firefighter Archer and Firefighter Walsh were killed when they were struck by pieces of an 18,000 gallon propane tank when the tank experienced a BLEVE. The piping leading from the tank was damaged when it was struck by an all terrain vehicle. A fire developed as a result of the leak and the fire department responded. While firefighters were protecting exposures, the tank exploded. Six other firefighters and a deputy sheriff were injured in the explosion. More information related to this incident is available in NIOSH Fire Fighter Fatality Investigation 98-F-14, report number 98-007-I-IA from the US Chemical Safety and Hazard Investigation Board, and from the National Fire Protection Association.

4/13/98

Michael Curtis Wiborg, Firefighter

Age 46, Paid-on-Call

Chanhassen Fire Department, Minnesota

Firefighter Wiborg died of a heart attack after completing an annual physical agility test/screening.

4/22/98

Ralph William Stanbery, Firefighter

Age 62, Volunteer

Granby Fire Department, Missouri

Firefighter Stanbery was assisting with the filling of brush trucks from a tanker (tender) at an arson wildland fire. He collapsed and subsequently died of a heart attack.

4/25/98
William J. Robertson, Battalion Chief
Age 46, Career
Ridge Road Fire District, New York

After exercising for 45 minutes on a treadmill at the fire station, Chief Robertson responded to a report of a car fire. When the incident proved to be in another jurisdiction, Chief Robertson began to return to quarters. He suffered a heart attack, his command vehicle left the roadway and struck a metal pole. The vehicle collision was observed by a security guard who rendered aid and was joined by a police officer and paramedics. The car fire was found to be arson.

4/29/98
Raymond Nakovics, Firefighter
Age 49, Career
New York City Fire Department, New York

Firefighter Nakovics suffered a heart attack at the scene of a multiple alarm highrise fire.

5/2/98
Joseph Kroboth, Jr., Fire Police Captain
Age 59, Volunteer
The Volunteer Fire Company of Halfway Maryland

Captain Kroboth was directing traffic at the scene of a serious motor vehicle collision. The scene was very busy and a medical helicopter was in the process of landing. A pickup truck suddenly changed lanes and struck Captain Kroboth, the driver's attention was directed toward incident operations. According to the police report, Kroboth was thrown 150 feet. Captain Kroboth was wearing a reflective vest and utilizing a flashlight with safety wand. Captain Kroboth died of massive head and chest injuries on 5/3/98.

5/5/98
Patrick Henry McKinney, Jr., Firefighter
Age 72, Volunteer
Colorado City Volunteer Fire Department, Texas

Firefighter McKinney was driving a converted tanker from one brush fire to another when he lost control of the tanker as it crossed a narrow bridge. The tanker rolled three times after the right rear wheels of the vehicle caught on a concrete guardrail. A firefighter who was a passenger in the tanker was injured in the collision. Firefighter McKinney was ejected.

5/7/98
Victor Clement Castillo, Fire Suppression Technician
Age 43, Career
El Paso Fire Department, Texas

Fire Suppression Technician Castillo was participating in a mandatory maze training exercise. During the event, Technician Castillo bumped his head twice, but told instructors that he was okay. After going off duty and going home, Technician Castillo became ill and was taken to the hospital by his wife. Later that month he was hospitalized for seizures and remained under a doctor's care until his death. He never returned to duty. The cause of death was ruled as cardiopulmonary arrest leading to anoxic brain injury. Underlying causes were aspiration secondary to a seizure, and seizure disorder secondary to the head injury that occurred in May of 1998. Further information related to this incident can be found in NIOSH Fire Fighter Fatality Investigation 99-F-08. Technician Castillo died on January 21, 1999.

5/9/98
Daniel W. Mumford, Firefighter/Driver
Age 48, Volunteer
West Haverstraw Fire Department, New York

Firefighter/Driver Mumford was the operator of a tower ladder apparatus at a mutual aid structure fire. While preparing to leave the scene, Firefighter/Driver Mumford was struck with an apparent heart attack. He had previously been under a doctor's care for a cardiac condition but had been released to drive a fire truck. Rain slicked surfaces and a fall by Firefighter/Driver Mumford may have contributed to his death.

5/19/98
Eugene Williard Blackmon, Jr., Firefighter
Age 38, Career
Chicago Fire Department, Illinois

Firefighter Blackmon was conducting an underwater search for two reported drowning victims in the Calumet River. While going from shore to a boat he lost his grip on a flotation device and slipped under the water. He had removed his SCUBA tank prior to entering the water. Firefighter Blackmon was recovered after approximately 10-15 minutes and provided with emergency medical care. He was airlifted by Fire Department helicopter to a local hospital but was pronounced dead at the hospital. Further information related to this incident can be found in NIOSH Fire Fighter Fatality Investigation 98-F-18.

5/26/98
Jake M. Hoeffner, Junior Fire Department Captain
Age 17, Volunteer
Yaphank Fire Department, New York

Firefighter Hoeffner was a passenger in the bed of a fire department pickup as it proceeded across a parking lot during preparation for fire department training. Firefighter Hoeffner fell from the pickup and struck his head, sustaining fatal injuries. He was not wearing a seat belt. Firefighter Hoeffner died on May 31, 1998.

5/27/98
Walter A. Ernst, Firefighter
Age 61, Volunteer
East Meadow Fire Department, New York

Firefighter Ernst assisted in the training of firefighters using SCBA's in a simulated smoke filled structure. He complained of pain in his left shoulder and was extremely fatigued at the conclusion of the training exercise as well as later in the fire station. He died at home in bed a few hours later of a heart attack.

5/29/98
Robert W. Munter, Fire Chief
Age 56, Career
Berlin Fire Department, Massachusetts

Chief Munter, the only paid member of his department, was conducting an inspection of a new school that was under construction in his jurisdiction. At some point during the inspection, Chief Munter suffered an unwitnessed medical emergency. He was found by construction workers lying prone with a head injury that likely occurred as he fell. Chief Munter was treated by members of his own Department and transported to the hospital, where he was pronounced dead. Chief Munter died of a heart attack.

6/2/98

Dennis L. Buroker, Sergeant

Age 44, Career

Muncie Fire Department, Indiana

Sergeant Buroker had gone to bed after responding to a brush fire at approximately 1:00 a.m. He was found dead in his bed by other firefighters in the morning. Sergeant Buroker had died in the night as the result of a heart attack due to hypertrophic cardiomyopathy (a genetic disease).

6/5/98

James Blackmore, Lieutenant, Age 48

Scott J. LaPiedra, Captain, Age 40

Career

New York City Fire Department, New York

Along with other firefighters, Lieutenant Blackmore and Captain LaPiedra were conducting a search on the second floor of a commercial/residential structure. A civilian fire victim had been reported to be trapped in the area. Without warning, the second floor collapsed into the fire area on the first floor, trapping firefighters in a live fire on the first floor. Two firefighters died and four were seriously injured. The civilian fire victim had escaped through a back entrance. Lieutenant Blackmore was pronounced dead at the hospital after being recovered by other firefighters, the cause of death was crushing trauma and burns resulting in a heart attack. Captain LaPiedra suffered severe burns (70%) and died on July 4, 1998, the cause of death was thermal burns resulting in cardiac arrest. More information related to this incident is available in NIOSH Fire Fighter Fatality Investigation 98-F-17.

6/26/98

Douglas L. Rohrbaugh, Fire Police Lieutenant

Age 52, Volunteer

Laurel Fire Company, Pennsylvania

Fire Police Lieutenant Rohrbaugh was directing traffic at the scene of a motor vehicle collision. He was struck from the rear by a pickup truck. The pickup left the scene without stopping. Lieutenant Rohrbaugh was thrown 97 feet and landed along the side of the road. EMS was provided at the scene but Lieutenant Rohrbaugh's injuries were too severe.

6/27/98
Jerry David Donahue, Pilot, Age 57
Charles Franklin Key, Copilot, Age 59
Wildland Contractor
Neptune Aviation Services, Missoula, Montana
Gila National Forest, New Mexico

Pilot Donahue and Copilot Key were killed as a result of the crash of their Lockheed SP-2H aircraft while fighting the Leggert wildland fire under contract for the United States Forest Service. The crash occurred about five miles West of Reserve, New Mexico. The tanker had completed a dry pass over the fire area, then circled around to make a second pass and release its load. At that time it contacted trees, crashed, and burned. The aircraft was carrying 2,450 gallons of fire retardant. More information related to this incident can be found in National Transportation Safety Board report FTW98GA86.

6/27/98
Johnnie Ray Park, Captain
Age 43, Career
Cullman Fire Department, Alabama

Captain Park died of a heart attack that occurred at the scene of a motor vehicle collision. The incident had been in progress for about an hour. Captain Park had assisted with patient treatment and scene cleanup. He was sitting in the cab of his engine beginning to complete an incident report when he was stricken. He was immediately transported to the hospital in the engine by his crew. The weather was hot and humid. Further information related to this incident can be found in NIOSH Fire Fighter Fatality Investigation 98-F-22.

6/29/98
Timothy D. Allen, Firefighter
Age 25, Volunteer
Central High Fire Department, Oklahoma

Firefighter Allen was responding to his fire station in his personal vehicle when he was involved in a collision with another vehicle at an uncontrolled intersection. Firefighter Allen was thrown through the windshield of his vehicle and landed almost twenty feet away. Two children that were passengers in Firefighter Allen's vehicle were injured. The driver of the other vehicle was injured. No one in either car was wearing a seat belt.

7/6/98
Tulon Lee Goodwin, Firefighter/Forestry Worker
Age 50, Wildland Career
Alabama Forestry Commission, Alabama

Firefighter Goodwin was stricken with a heart attack at the scene of a wildland fire. The fire was caused by children playing with bottle rockets. Firefighter Goodwin had been operating on the scene for about four hours. He had been plowing a fire line.

7/17/98
John Rochford Kennedy, Fire Police Officer
Age 75, Volunteer
Ocean Pines Volunteer Fire Department, Maryland

Fire Police Officer Kennedy was stricken with a heart attack as he directed traffic at the scene of a motor vehicle collision.

7/23/98
Matthew P. Casboni, Firefighter
Age 55, Volunteer
Saint John Volunteer Fire Department, Indiana

Firefighter Casboni died of a heart attack that occurred as he was acting as the air supply officer at a working structure fire.

7/23/98
Thomas E. Prendergast, Captain
Age 56, Career
Chicago Fire Department, Illinois

Captain Prendergast and his crew were fighting a two alarm fire in a residential occupancy. Captain Prendergast and his crew were operating hose lines when he complained of chest pain and shortness of breath. He was immediately escorted to an ambulance and transported to the hospital. Captain Prendergast died as a result of a heart attack. He died on August 8, 1998.

8/1/98

Barvon Coy Hamilton, Firefighter

Age 71, Volunteer

Southern Oaks Volunteer Fire Department, Texas

Firefighter Hamilton was on the back step of a pumper as it was relocated at the scene of a brush fire. The pumper was backing up, Firefighter Hamilton was on the back step to secure hose that had been reloaded. As the pumper backed up, Firefighter Hamilton attempted to warn the driver about a utility pole that was behind the apparatus. He apparently lost his grip and was crushed between the pumper and the pole. His right leg was amputated below the knee. Despite the efforts of local EMS providers, Firefighter Hamilton died of hypovolemic shock (loss of blood). Further information related to this incident can be found in NIOSH Fire Fighter Fatality Investigation 98-F-19.

8/3/98

Donald Claude Martin, Firefighter

Age 34, Volunteer

Van Buren Fire Department, Maine

Firefighter Martin was in his fire station preparing to respond to assist with a search for a missing child. He became ill and was transported by other firefighters to the hospital. He was pronounced dead upon arrival. Firefighter Martin died of a heart attack.

8/16/98

Larry Joe King, Firefighter

Age 42, Paid-Call

Maury City Fire Department, Tennessee

Firefighter King was attempting to pry open the hood of a pickup truck that was on fire. He suffered a heart attack and died.

8/18/98

Calvin Harbaugh, Sr., Fire Police Officer

Age 56, Volunteer

Ebenezer Fire Company, Pennsylvania

Fire Police Officer Harbaugh was stricken with a heart attack while directing traffic at the scene of a motor vehicle collision.

8/20/98

John M. Walker, Private

Age 58, Career

Memphis Fire Department, Tennessee

Private Walker was assigned on light duty to the Department's SCBA maintenance shop. He was stricken with a heart attack while on duty. Despite immediate ALS care, Private Walker died.

8/29/98

Justin Melton, Firefighter, Age 22

Scott Selby, Firefighter, Age 35

Volunteer

Marks Fire Department, Mississippi

Firefighters Melton and Selby were working in different areas of a structure fire that involved a commercial building. A collapse occurred which trapped Firefighter Melton as he and other firefighters were advancing a hoseline on the fire. Firefighter Selby was on the roof of the fire structure attempting ventilation when he fell into the fire area and was killed. Both firefighters died of asphyxiation due to smoke inhalation. Further information related to this incident can be found in NIOSH Fire Fighter Fatality Investigation 98-F-21.

8/29/98

Robert F. Peters, Lieutenant

Age 52, Volunteer

Hastings on Hudson Fire Department, New York

Lieutenant Peters was completing paperwork after returning from a response to an automatic fire alarm. Lieutenant Peters had driven an aerial apparatus to the incident. He was stricken with a heart attack and died.

8/31/98

Brian Carrasco, Inmate Handcrew Firefighter

Age 35, Wildland Part-Time

Los Angeles County Fire Department, California

Firefighter Carrasco was killed in a vehicle collision while working as a member of an inmate handcrew. The brush/engine vehicle in which we was riding rolled over. Eleven others were injured.

9/4/98
Juan Manuel Hernandez, Jr., Firefighter
Age 21, Wildland Full-Time Seasonal
United States Department of Agriculture Forest Service, New Mexico

Firefighter Hernandez was killed in a vehicle collision while working near Willows, California. The engine in which he was riding was struck by a pickup truck that crossed the center line and impacted the engine along the left underside and rear dual wheels. The engine flipped and landed upside down. Firefighter Hernandez was wearing a seat belt but was partially ejected from the vehicle. He was pronounced dead at the scene.

9/4/98
Allen Lawrence Heirtzler, Firefighter
Age 22, Volunteer
Slaughter Volunteer Fire Department, Louisiana

Firefighter Heirtzler was responding to a trailer fire in his role as a volunteer firefighter. He was responding is his Sheriffs Department cruiser when he was involved in a collision. Firefighter Heirtzler's vehicle was traveling at a high rate of speed when it began a skid, the vehicle struck a cow in the roadway, then crashed into a utility pole, rolled over, and burned. Firefighter Heirtzler was not wearing a seat belt at the time of the collision.

9/5/98
Eugene P. McDonough, Firefighter
Age 54, Career
Saint Johnsbury Fire Department, Vermont

Firefighter McDonough responded with other members of his Department to a mutual aid fire in a recycling facility. While opening a large door to allow a master stream attack, Firefighter McDonough was crushed when a parapet wall collapsed. The cause of the fire was arson. Further information related to this incident can be found in NIOSH Fire Fighter Fatality Investigation 98-F-20.

9/9/98
Ernest Alan McElroy, Forest Ranger II
Age 40, Wildland Career
Arkansas Forestry Commission, Arkansas

Forest Ranger McElroy was plowing a fire line with a bulldozer. The fire overcame his position, he attempted to back out but struck a tree. Ranger McElroy dismounted the bulldozer and proceeded down the fire line on foot. He was burned over 60% of his body but still managed to walk the half mile to a waiting ambulance. Ranger McElroy died of his injuries on October 28,

1998. Further information related to this incident can be found in NIOSH Fire Fighter Fatality Investigation 98-F-30.

9/14/98
Randy Sims, Captain
Age 44, Volunteer
Antioch Volunteer Fire Department, South Carolina

Captain Sims was at the scene of a structure fire assisting with overhaul. He was stricken with a heart attack and died.

9/18/98
Donald Trotochaud, Firefighter-Senior Airman
Age 23, Career
Laughlin Air Force Base, Texas

Firefighter Trotochaud was killed in a single vehicle collision while enroute to standby at a remote airfield. One other firefighter was injured.

9/21/98
David M. Brinkley, Firefighter – Past Chief
Age 43, Volunteer
United Communities Volunteer Fire Department, Maryland

Firefighter Brinkley was stricken with a heart attack while refilling SCBA cylinders after a response to a vehicle fire.

9/24/99
Tony B. Chapin, Firefighter/EMT
Age 19, Volunteer
Willamina Fire Department, Oregon

Firefighter Chapin was killed while on the way to a paramedic training class in his personal vehicle. A car crossed the center line and struck the vehicle that Firefighter Chapin was driving. He survived the initial impact but died the next day. Firefighter Chapin was wearing his seat belt.

9/27/98
Preston Edgar Patterson, Firefighter – Fire Police
Age 66, Volunteer
The Manchester Fire Engine and Hook and Ladder Company Number One, Maryland

Firefighter Patterson was stricken with a heart attack while performing Fire Police duties at the scene of a motor vehicle collision.

9/28/98
Neil A. Holmes, Captain
Age 55, Career
Fresno City Fire Department, California

Captain Holmes was found unconscious in the restroom of his fire station. He had succumbed to a brain aneurysm.

9/28/98
Paul P. Satterfield, Battalion Chief
Age 60, Career
Nashville Fire Department, Tennessee

Chief Satterfield was in command of a fire on September 28[th.] He complained of illness at the fire and went off duty in the morning. He was found dead at home. He had died of a brain aneurysm. Chief Satterfield died on 9/29/98.

9/30/98
Robert Odell Lee, Fire Chief
Age 56, Volunteer
North River Valley Volunteer Fire Company, West Virginia

Chief Lee was monitoring pump operations during a drill at a nursing home. He was wearing full protective clothing but no SCBA. He was not performing any strenuous activity, however the day was hot and humid. Chief Lee suddenly collapsed and died of a heart attack. Further information related to this incident can be found in NIOSH Fire Fighter Fatality Investigation 98-F-11.

10/5/98
Gary D. Nagel, Airtanker Pilot
Age 62, Wildland Contractor
San Joaquin Helicopters, California

Airtanker Pilot Nagel was killed in the crash of a Grumman TS-2A airtanker when he misjudged his maneuvering altitude and impacted the terrain. He had made two drops on the Mount Edna fire near Banning, California and was preparing to make a third. Other factors that contributed to the crash were the mountainous terrain, tailwind conditions, and turbulence in the area. Airtanker Pilot Nagel was an employee of San Joaquin Helicopters, a contractor to the California Department of Forestry and Fire Protection. More information related to this incident can be found in National Transportation Safety Board report LAX99GA005.

10/5/98
Thomas Oscar Wall, Captain
Age 44, Career
Orange County Fire Authority, California

Captain Wall died at the Taylor fire in Riverside, California. He was protecting exposed dwellings when he told other firefighters that he did not feel well and collapsed. Captain Wall died of a heart attack.

10/13/98
Barry L. Wary, Firefighter
Age 51, Volunteer
Klingerstown Volunteer Fire Company, Pennsylvania

Firefighter Wary was actively involved in the suppression of a fire in an industrial occupancy. Upon exiting the structure, he collapsed and died of a heart attack.

10/24/98
Carson L. Gosey, Sr., Firefighter – Fire Police
Age 60, Volunteer
Shiloh/Danieltown/Oakland Fire Department, North Carolina

Firefighter Gosey was struck by a vehicle at the scene of a training exercise as he assisted a water tanker that was crossing the road.

10/24/98
Lawrence D. Thrower, Lieutenant
Age 51, Volunteer
Sidney Fire Department, New York

Lieutenant Thrower responded to the scene of a dumpster fire at a manufacturing facility. He was equipped in full protective clothing and SCBA as he and his crew extinguished and overhauled the fire. Lieutenant Thrower had removed his facepiece at the conclusion of operations and was beginning to remove his other protective clothing when he collapsed. ALS was administered at the scene and he was transported to the hospital. Lieutenant Thrower was pronounced dead of a heart attack a short time later.

11/6/98
Hubert Sidney Jones, Fire Chief
Age 29, Volunteer
Thoroughfare Volunteer Fire Department, North Carolina

Robby Dean Blizzard, First Lieutenant
Age 24, Volunteer
Arrington County Volunteer Fire Department

Chief Jones and First Lieutenant Blizzard were killed as they fought a fire in an automobile salvage yard storage building. Firefighters believed that they had found the seat of the fire and were applying water when a rapid change in conditions occurred. Chief Jones ran out of air while trying to escape. Lieutenant Blizzard entered the structure to search for Chief Jones. He ran out of air, became disoriented, and failed to exit the building. Lieutenant Blizzard was wearing a PASS device but it was not activated. Chief Jones was not equipped with a PASS device. The cause of death for Chief Jones was listed as carbon monoxide poisoning and smoke inhalation and the cause of death for Lieutenant Blizzard was listed as carbon monoxide poisoning. Lieutenant Blizzard was also a career firefighter in another community but was off duty at the time. Further information related to this incident can be found in NIOSH Fire Fighter Fatality Investigation 98-F-32.

11/7/98
Paul Allen Laux, Captain
Age 50, Volunteer
Cool Spring Township Fire Department, Indiana

Captain Laux drove a water tanker to the scene of a reported structure fire. When the report turned out to be steam, the incident commander released all units to return to station. As reports were completed and units prepared to return to station, a firefighter noticed that the door to the tanker was open. As he looked inside, he observed Captain Laux slumped over the wheel and unconscious. Captain Laux was removed from the tanker and provided with medical care, including the arrival of ALS. No defibrillator was available initially but one was utilized upon the arrival of ALS. Captain Laux had a history of cardiovascular disease including bypass surgery but had been cleared for fire fighting by his personal physician. Further information related to this incident can be found in NIOSH Fire Fighter Fatality Investigation 99-F-05.

11/8/98

Charles Peter Frank III, Deputy Chief

Age 56, Volunteer

West Weatherfield Fire Department, Vermont

Deputy Chief Frank was in command of a vehicle fire on a local highway. He participated in forcible entry, water supply, and fire attack. After the fire was extinguished, Chief Frank began to speak with the people who reported the fire when he suddenly collapsed. Despite immediate EMS assistance on the scene and the arrival of ALS, Chief Frank died of a heart attack.

11/9/98

Thomas Benjamin Rice, Fire Police Officer

Age 70, Volunteer

Village of Perry Fire Department, New York

Fire Police Officer Rice had been directing traffic at the scene of a commercial structure fire. His duties were completed and he had been released. A passing firefighter noticed Officer Rice on the ground near his vehicle. Despite immediate aid, he died of a heart attack. Nothing unusual preceded the attack. Officer Rice had a history of heart problems.

11/13/98

William Dwight Yankey, Firefighter

Age 35, Career

Harrodsburg Fire Department, Kentucky

Firefighter Yankey's Captain was awakened when Firefighter Yankey fell out of bed. His Captain found him not breathing and began CPR with the assistance of other firefighters. Firefighter Yankey regained consciousness at least twice and spoke with other firefighters. He was transported to the hospital and was pronounced dead. The cause of death was listed as acute cardiorespiratory failure due to occlusive pulmonary thromboemboli (blood clot in the lungs).

11/18/98

Roger DeWayne Bookout, Heavy Fire Equipment Operator

Age Unknown, Wildland Career

California Department of Forestry and Fire Protection, California

Heavy equipment operator Bookout was killed in an unwitnessed rollover of the tractor that he was using to perform fire lookout road maintenance.

11/24/98
Norman Neal Almond, Captain, Age 46
Craig Daniel Brown, Driver/Operator, Age 27
Career
Parsons Fire Department, Kansas

A painting contractor found that their ladders were not long enough to reach the upper portion of the church. A representative of the church contacted city hall and requested an aerial ladder. After an assessment by the fire chief, a reserve pumper was sent to the scene to allow for the use of its ground ladders. Captain Almond and Driver/Operator Brown were assisting with the positioning of a ladder on the exterior of a church that was being painted. The aluminum ladder made contact with an electrical service line, resulting in the fatal electrocution of both firefighters and injury to one additional firefighter. The firefighters were positioning the ladder since it was too cumbersome for the two painters to position by themselves. Further information related to this incident can be found in NIOSH Fire Fighter Fatality Investigation 98-F-31.

12/2/98
Steven C. Mayfield, Firefighter
Age 47, Career
Houston Fire Department, Texas

Firefighter Mayfield was participating in Federal Aviation Administration mandated Aircraft Rescue Fire Fighting (ARFF) training at the Dallas-Fort Worth airport training facility. Firefighter Mayfield was in the interior of an aircraft fuselage mockup lifting and pulling a mannequin when he experienced a fatal heart attack.

12/9/98
Steve Austin Tippins, Assistant Chief
Age 37, Volunteer
Etolie Volunteer Fire Department, Texas

Chief Tippins was responding in his personal vehicle to an EMS incident. He failed to yield at a stop sign and was broadsided on the driver's side by a truck. Chief Tippins was killed instantly. His wife, as passenger in his vehicle, was injured.

12/12/98
Stephen E. Gessler, First Assistant Chief
Age 44, Volunteer
Little Falls Fire Department, New Jersey

Chief Gessler was in command of a search effort to find a missing civilian. The man was located and Chief Gessler ordered an ambulance to transport him. Immediately after giving the order, Chief Gessler complained of being dizzy and collapsed. Immediate EMS efforts were begun but were not successful in saving Chief Gessler. He died of dilated cardiomyopathy (a disorder in which the heart muscle is weakened and cannot pump blood efficiently).

12/18/98
James F. Bohan, Firefighter, Age 25
Christopher M. Bopp, Firefighter, Age 27
Joseph P. Cavalieri, Lieutenant, Age 42
Career
New York City Fire Department, New York

Firefighter Bohan, Firefighter Bopp, and Lieutenant Cavalieri were killed while fighting a residential high rise structure fire. As they rushed to the tenth floor to search for victims, they were overcome by a wave of heat and smoke that killed all three. The heat wave, or fireball, may have been propelled by a gust of wind coming through the fire apartment. The automatic closing device on the apartment door had been removed or had malfunctioned. The building's hallway sprinklers did not activate due to a closed valve. Six firefighters were injured in the fire.

12/18/98
Thomas J. Concannon, Fire Police Lieutenant
Age 55, Volunteer
Wormleysburg Fire Company #1, Pennsylvania

Fire Police Lieutenant Concannon responded in his duties as a Fire Police member to a motor vehicle collision. He parked his vehicle short of the incident scene. Members working on the incident scene assumed that he had stopped in that location to divert traffic. At the conclusion of the incident, Lieutenant Concannon failed to respond to a radio call. Shortly thereafter, he was found slumped over the steering wheel of his vehicle, unresponsive, with the vehicle still in gear and his foot on the brake. Despite immediate medical care, Lieutenant Concannon died of a heart attack. Further information related to this incident can be found in NIOSH Fire Fighter Fatality Investigation 99-F-10.

12/31/98
Kennon Loy Williams, Captain
Age 27, Volunteer
Banks County Fire Department, Georgia

Captain Williams and other members of his Department were conducting an offensive attack on a arson fire of a church built around 1850. Captain Williams was caught under heavy timbers in a roof collapse. Further information related to this incident can be found in NIOSH Fire Fighter Fatality Investigation 99-F-04.

The following firefighters died in 1998. They are not included in the analysis since one died as the result of an injury sustained in a previous year and two worked and died outside of the fifty states and the District of Columbia that have historically been included in this analysis.

7/11/89
Walter Bitner, Firefighter
Age 38, Career
Paterson Fire Department, New Jersey

Firefighter Bitner died on October 24, 1998, as a result of injuries he received on July 11, 1989. Firefighter Bitner was responding to what turned out to be a false alarm when he fell from the jumpseat of his engine company and sustained severe head injuries. Firefighter Bitner was in a coma for six months. He remained under medical care for the rest of his life, unable to perform any physical task other than moving his hands and his eyes.

2/5/98
Luis A. Rivera-Rivas, Firefighter
Age 58, Career
Puerto Rico Fire Department, Puerto Rico

Died of a heart attack after being exposed to smoke at a grass fire in the town of Papillas. The wind shifted resulting in the smoke exposure for Firefighter Rivera-Rivas.

5/7/98
Jesus Mercado, Firefighter
Age 35, Career
Puerto Rico Fire Department, Puerto Rico

Firefighter Mercado was killed in the rollover of a ladder truck while enroute to a furniture warehouse fire in the town of Bayamon. One of the rear tires on the apparatus failed, resulting in the collision. Four other firefighters were injured.

Additional information about many of the firefighter fatalities presented in this appendix is available from the sources below. Where known, the report number for each incident is listed in the appendix along with the incident description. Many reports are available through the mail and the internet.

National Fire Protection Association
1 Batterymarch Park
P.O. Box 9101
Quincy, MA 02269
(617)-770-3000
http://www.nfpa.org

National Institute for Occupational Safety and Health (NIOSH)
Fire Fighter Fatality Investigation and Prevention Program
1095 Willowdale Road
Mail Stop P-180
Morgantown, WV 26505-2888
http://www.cdc.gov/niosh/firehome.html
(800)-35-NIOSH

National Transportation Safety Board
http://www.ntsb.gov/NTSB/Query.htm

US Chemical Safety and Hazard Investigation Board
Office of External Relations
2175 K Street, Northwest
Washington, DC 20037
(202)-261-7600
http://www.chemsafety.gov

www.ingramcontent.com/pod-product-compliance
Lightning Source LLC
Chambersburg PA
CBHW081222170526
45165CB00009B/2913